Middle Mississippi River Decision Support System: User's Manual

By Jason J. Rohweder, Steven J. Zigler, Timothy J. Fox, and Steven N. Hulse

Chapter 1 of
Book 6, Modeling Techniques, Section C, Decision Support Systems

Techniques and Methods 6 C–1

U.S. Department of the Interior
U.S. Geological Survey

U.S. Department of the Interior
Gale A. Norton, Secretary

U.S. Geological Survey
Charles G. Groat, Director

U.S. Geological Survey, Reston, Virginia: 2005

For more information about the USGS and its products:
Telephone: 1-888-ASK-USGS
World Wide Web: http://www.usgs.gov/

Suggested citation:
Rohweder, J. J., S. J. Zigler, S. N. Hulse, and T. J. Fox, 2005, Techniques and Methods Book 6, Modeling Techniques, Section C, Decision Support Systems, Chapter 1, Middle Mississippi River Decision Support System: User's Manual: U.S. Geological Survey, Techniques and Methods 6 C–1, 53 p. + CD -ROM.

Contents

Figures

Middle Mississippi River Decision Support System: User's Manual

By Jason J. Rohweder,[1] Steven J. Zigler,[1] Timothy J. Fox,[1] and Steven N. Hulse[2]

Abstract

This user's manual describes the Middle Mississippi River Decision Support System (MMRDSS) and gives detailed examples on its use. The MMRDSS provides a framework to assist decision makers regarding natural resource issues in the Middle Mississippi River floodplain. The MMRDSS is designed to provide users with a spatially explicit tool for tasks, such as inventorying existing knowledge, developing models to investigate the potential effects of management decisions, generating hypotheses to advance scientific understanding, and developing scientifically defensible studies and monitoring. The MMRDSS also includes advanced tools to assist users in evaluating differences in complexity, connectivity, and structure of aquatic habitats among river reaches. The Environmental Systems Research Institute ArcView 3.x platform was used to create and package the data and tools of the MMRDSS.

Introduction

The Middle Mississippi River is described as the section of the Mississippi River between the confluence of the Missouri and Ohio Rivers. The Middle Mississippi River and its floodplain have been substantially modified to accommodate multiple human uses that include urban development, navigation, and agriculture. Consequently, policy makers in state and Federal agencies frequently face difficult decisions regarding management of biological resources in the river. The Middle Mississippi River Decision Support System (MMRDSS) provides a framework to assist decision makers regarding natural resource issues in the Middle Mississippi River floodplain. The MMRDSS is designed to provide users with a spatially explicit tool for tasks, such as inventorying existing knowledge, developing models to investigate the potential effects of management decisions, generating hypotheses to advance

scientific understanding, and developing scientifically defensible studies and monitoring. To facilitate development of the Decision Support System (DSS), an interagency workshop was held in Cape Girardeau, Missouri, in February 2002. The workshop focused on hydrologic and other system drivers of aquatic habitats and their role structuring aquatic habitats and habitat connectivity. Workshop participants identified geographic information system (GIS) data themes that might be useful for addressing key natural resource issues in the Middle Mississippi River floodplain such as pallid sturgeon (*Scaphirhynchus albus*), habitat restoration, floodplain connectivity, and ecologically meaningful classification systems for large river habitats.

Decision support systems are intended to assist managers and researchers in their decision processes by facilitating semi-structured or unstructured tasks. GIS-based DSSs can provide a common platform for gathering spatial data from disparate state, Federal, and nongovernment sources. They often incorporate tools to enable "on-the-fly" tasks, such as mapping, queries, quantification of data elements, modeling, and output.

ArcView GIS Platform

A GIS is a tool that allows the user to analyze, display, create, edit, and output geographically referenced data. Environmental System Research Institute's (ESRI; Redlands, CA) ArcView 3.x software was used as the platform for the MMRDSS. ArcView was selected because of its relative ease-of-use, easily customized interface, powerful analytical tools, low cost, and widespread availability.

Discussion

The MMRDSS is a modified ArcView 3.x GIS project (fig. 1). Within the MMRDSS, the user has access to a variety of spatial data themes corresponding to the Middle Mississippi River. The ArcView project has also been programmed to include several advanced tools to assist the user in their decision making.

[1]U.S. Geological Survey, Upper Midwest Environmental Sciences Center, 2630 Fanta Reed Road, La Crosse, Wisconsin 54603

[2]U.S. Geological Survey, Department of Fisheries and Wildlife Services, Cooperative Research Unit, 302 ABNR Building, University of Missouri, Columbia, Missouri 65211

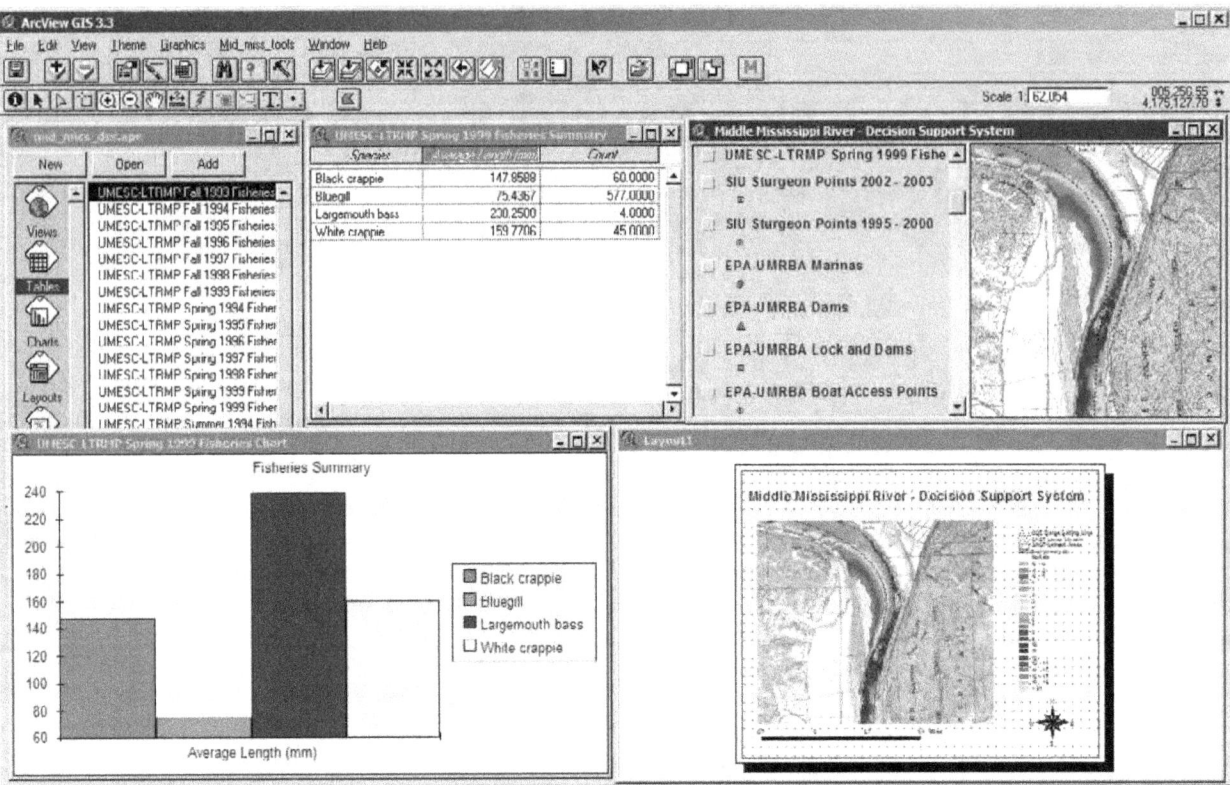

Figure 1. ArcView 3.x GIS interface.

The User's Manual for the MMRDSS is provided in Appendix 1. Appendix 2 lists GIS data themes used in the MMRDSS. Appendix 3 lists the metrics calculated for aquatic area classifications and bathymetry using FRAGSTATS Version 3.3 (McGarigal et al. 2002).

The MMRDSS was developed with the ESRI Avenue macro-language and can be used with any ArcView version 3.x. Output from the tools can be further enhanced by using ArcView's GIS tools and by incorporating user data. **The MMRDSS will not work within ArcView versions 8.x and 9.x, or ArcGIS.**

Acknowledgments

We thank Carl Korschgen, Douglas Olsen, Mark Laustrup, Robert Jacobson, Linda Leake, Barry Johnson, Kirk Lohman, and Georginia Ardinger for their thoughtful comments on various versions of this manual. Robert Hrabik, Valerie Barko, and Carl Korschgen also provided helpful guidance for developing the Middle Mississippi River Decision Support System.

References

Amoros, C., and G. Bornette, 2002, Connectivity and biocomplexity in waterbodies of riverine floodplains: Freshwater Biology, v. 47, p. 761–776.

Awasthi, K. D., B. K. Sitaula, B. R. Singh, and R. M. Bajacharaya, 2002, Land-use change in two Nepalese watersheds: GIS and geomorphometric analysis: Land Degradation and Development, v. 13, p. 495–513.

DeHaan, H. C., T. J. Fox, C. E. Korschgen, C. H. Theiling, and J. J. Rohweder, 2000, Habitat Needs Assessment GIS Query Tool for the Upper Mississippi River System: User's Manual: La Crosse, Wis. Prepared for the U.S. Army Corps of Engineers, St. Louis District, Missouri.

Fausch, K. D., C. E. Torgersen, C. V. Baxter, and H.W. Li, 2002, Landscapes to riverscapes: bridging the gap between research and conservation of stream fishes: BioScience, v. 52, p. 483–498.

Forman, R. T. T., and M. Godron, 1984, Landscape ecology: New York, Wiley.

Haltuch, M. A., and P. A. Berkman, 2000, Geographic information system (GIS) analysis of ecosystem invasion: exotic mussels in Lake Erie: Limnology and Oceanography, v. 45, p. 1778–1787.

Harlin, J. M., 1978, Statistical moments of the hypsometric curve and its density function: Mathematical Geology, v. 10, p. 59–72.

Hulse, S. N., and D. L. Galat, 2003, Middle Mississippi River Decision Support System Year End Project Progress Report and Landscape Attributes Preliminary Product Release: Columbia, Mo., U.S. Geological Survey, Missouri Cooperative Fish and Wildlife Unit report, prepared for the U.S. Geological Survey, Upper Midwest Environmental Sciences Center, La Crosse, Wisconsin.

Hurtrez, J.–E., F. Lucazeau, J. Lave, and J.–P. Avouac, 1999, Investigation of the relationships between basin morphology, tectonic uplift, and denudation from the study of an active fold belt in the Siwalik Hills, central Nepal: Journal of Geophysical Research, v. 104, B6, p. 12779–12796.

Hurtrez, J.–E., and F. Lucazeau, 1999, Effect of drainage area on hypsometry from an analysis of small-scale drainage basins in the Siwalik Hills (Central Nepal): Earth Surface Processes and Landforms, v. 24, p. 799–808.

Johnson, B. L., W. B. Richardson, and T. J. Naimo, 1995, Past, present, and future concepts in large river ecology: BioScience, v. 45, p. 134–141.

Koel, T. M., 2004, Spatial variation in fish species richness of the Upper Mississippi River System: Transactions of the American Fisheries Society, v. 133, p. 984–1003.

Lee, J., and D. W. S. Wong, 2001, Statistical analysis with ArcView GIS: New York, Wiley.

Legendre, P., and L. Legendre, 1998, Numerical ecology: Amsterdam, The Netherlands, Elsevier Science B.V.

Leopold, L. B., M. G . Wolman, and J. P. Miller, 1964, Fluvial processes in geomorphology: San Francisco, W.H. Freeman and Company.

Lo, C. P., and A. K. W. Yeung, 2002, Concepts and techniques of geographic information systems: Upper Saddle River, N. J., Prentice Hall.

Luo, W., 1998, Hypsometric analysis with a geographic information system: Computers and Geosciences, v. 24, p. 815–821.

McGarigal, K., S. A. Cushman, M. C. Neel, and E. Ene, 2002, FRAGSTATS: spatial pattern analysis program for categorical maps. Computer software program produced by the authors at the University of Massachusetts, Amherst. Available at the following web site: www.umass.edu/landeco/research/fragstats/fragstats.html.

Wildhaber, M. L., P. J. Lamberson, and D. L. Galat, 2003, A comparison of measures of riverbed form for evaluating distributions of benthic fishes: North American Journal of Fisheries Management, v. 23, p. 543–557.

Wlosinski, J. H., and L. B. Wlosinski, 2001, Predicting flood potential to assist reforestation for the Upper Mississippi River System: La Crosse, Wis., U.S. Geological Survey, Project Status Report PSR 2001–01. Available at the following web site: http://www.umesc.usgs.gov/data_library/water_elevation/flood_potential.html.

Appendix 1: User's Manual

Section 1: General Information

Minimum System Requirements

The minimum system requirements include a personal computer with a Pentium processor, 32 MB random access memory, Microsoft Windows 95/98/2000/NT/XP, 1.2 GB hard drive space, and Environmental Systems Research Institute's (ESRI; Redlands, CA) ArcView 3.x software. Users must have a basic knowledge of computers and ESRI's ArcView program. **This software will not work within ArcView versions 8.x and 9.x, or ArcGIS.**

ArcView GIS Platform

A Geographic Information System (GIS) is a tool that allows the user to analyze, display, create, edit, and output geographically referenced data. The ESRI's ArcView 3.x software was used as the platform for the Middle Mississippi River Decision Support System (MMRDSS). ArcView was selected because of its relative ease-of-use, easily customized interface, powerful analytical tools, low cost, and widespread availability.

Using the Manual (Hard Copy/Digital)

This manual was developed to assist people in using the MMRDSS with the assumption that users have a basic knowledge of computers and ESRI's ArcView 3.x GIS software. The manual is similar in structure to other software user's manuals available today. Throughout the manual, certain conventions have been applied. Field names are in ***bold and italics***, file names are in *italics*, and features that are to be selected by the user are in **bold**. Section 1 of the manual provides an introduction and information (e.g., system requirements) needed before installing the tools, section 2 details the tool installation process, section 3 reviews the data of the MMRDSS, section 4 discusses several basic ArcView functions, and sections 5 and 6 describe the advanced geospatial and analysis tools included within the MMRDSS.

A digital copy of the manual is also available in PDF format in the *\mid_miss_dss\user_manual* directory on MMRDSS–CD1 and is readable by Adobe Acrobat Reader. This digital version has hypertext that allows the user to navigate to the sections of the manual.

ArcView's online help (fig. 1–1) is also available to answer questions about ArcView 3.x and its analytical tools.

Technical Support

This manual and ArcView's online help should address the majority of the user's questions. For additional help, contact the following:

U. S. Geological Survey
Upper Midwest Environmental Sciences Center
Geospatial Sciences and Decision Support Laboratory
2630 Fanta Reed Road
La Crosse, Wisconsin 54603
Telephone: (608) 783–6451

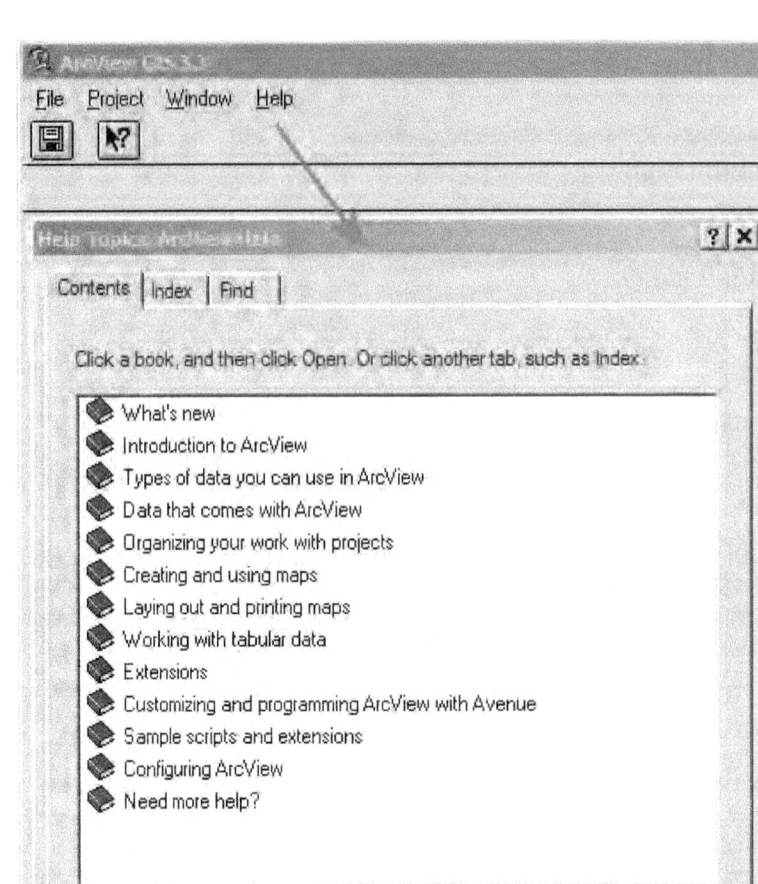

Figure 1–1.   ArcView's online help window.

Section 2: Getting Started

Installing the Middle Mississippi River Decision Support System

Important: Before the MMRDSS is installed, ensure that 1.2 GB of free space are available on the hard drive to copy all the tools and data from the CD-ROMs. The CD-ROMs are attached to the inside of the back cover of this manual.

1. Insert the MMRDSS–CD1 into the computer's CD-ROM drive.
2. Open Windows Explorer on the computer by clicking on the **Start** button, navigate to **Programs -> Accessories -> Windows Explorer**.
3. In Windows Explorer, **double-click** on the **CD-ROM drive** on the computer.
4. Copy the entire directory *mid_miss_dss*\\ (548 MB) from the MMRDSS–CD1 to the computer at the root level of any hard drive (e.g., C:\\fig. 2–1). **Since the ArcView project links to data copied from the CD-ROM, it is imperative that you copy the *mid_miss_dss*\\ directory to the root level and not into another existing directory!**
5. Once the files are copied, insert the MMRDSS–CD2 into the computer's CD-ROM drive.
6 In Windows Explorer, **double-click** on the **CD-ROM drive** on the computer.
7. Copy the entire directory *shapefiles*\\ (617 MB) from the MMRDSS–CD2 to the directory *mid_miss_dss**data*\\ that was copied previously (fig. 2–2).

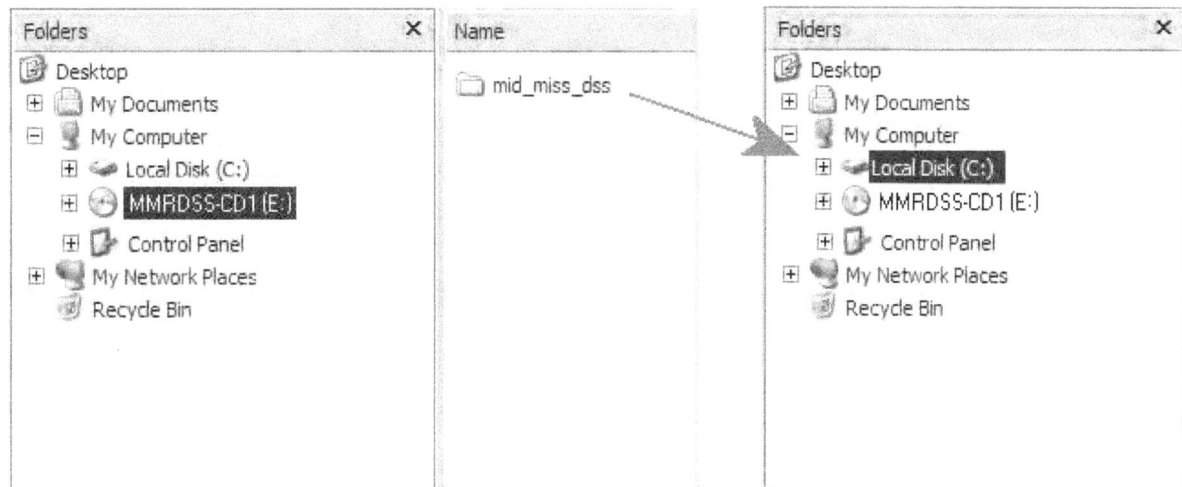

Figure 2–1. Copying the MMRDSS–CD1 data to the hard drive.

Copying MrSid Files to the Hard Drive

1. Two files have been included with the MMRDSS that allow the user to view compressed image files (MrSid; Lizardtech; Seattle, WA) within ArcView. Copy these files to the location on the hard drive where you have installed ArcView.
2. First, copy the file *mrsid.avx* to the *EXT32*\\ directory. If ArcView is installed on the C drive, there will be a directory *C:\\ESRI*\\. If the directory is not there, check the other possibilities on the computer (e.g., *D:\\ESRI*\\). Once you have located the *ESRI*\\ directory, navigate to the directory *ESRI**AV_GIS30**ARCVIEW**EXT32*\\.
3. Next, copy the file *mid_miss_dss**data**misc**MrSid_files**mrsid.avx* to this directory (*ESRI**AV_GIS30**ARCVIEW**EXT32*\\; fig. 2–3).
4. If a window opens and contains the text, "This folder already contains a file named 'mrsid.avx', would you like to replace the existing file," check to see if the existing file displayed is older than the file you are copying. If the existing file is older, click on the **Yes** button, if not, click on the **No** button.
5. Next, copy the file *mid_miss_dss**data**misc**MrSid_files**AVMrSID.dll*\\ to the directory *ESRI**AV_GIS30**ARCVIEW**BIN32*\\ (fig. 2–3).
6. If a window opens and contains the text "This folder already contains a file named 'avmrsid.dll', would you like to replace the existing file" check to see if the existing file displayed is older than the file you are copying. If the existing file is older, click on the **Yes** button, if not, click on the **No** button.
7. Once these files are copied, you will be able to use the MMRDSS.

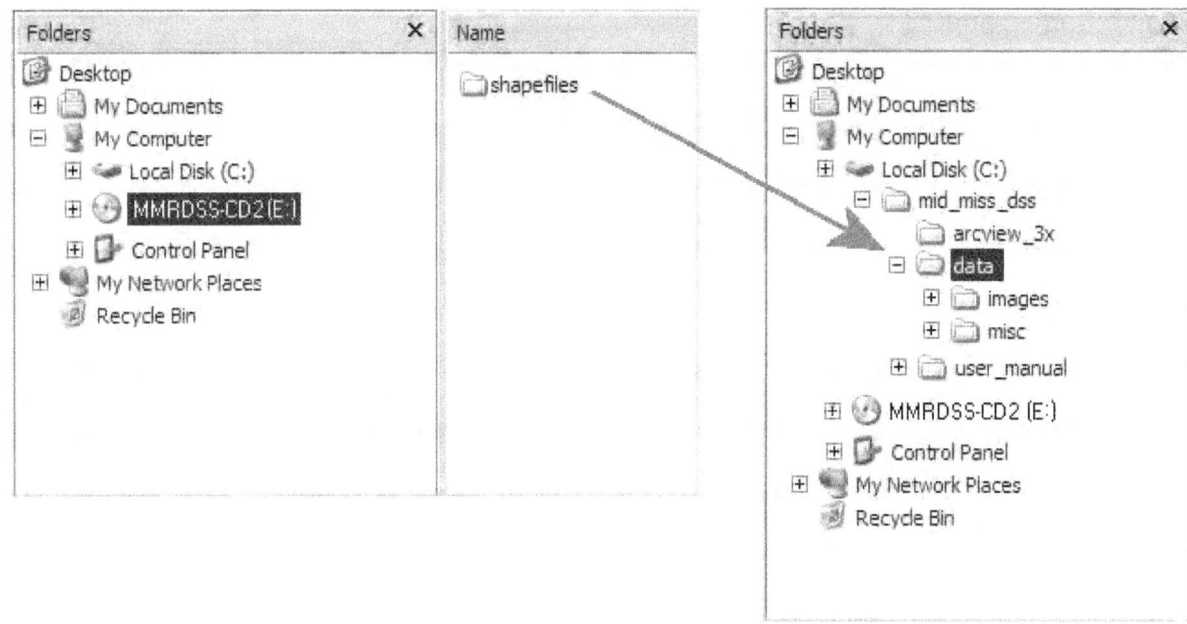

Figure 2–2. Copying the MMRDSS–CD2 *shapefiles* directory to the *data* directory.

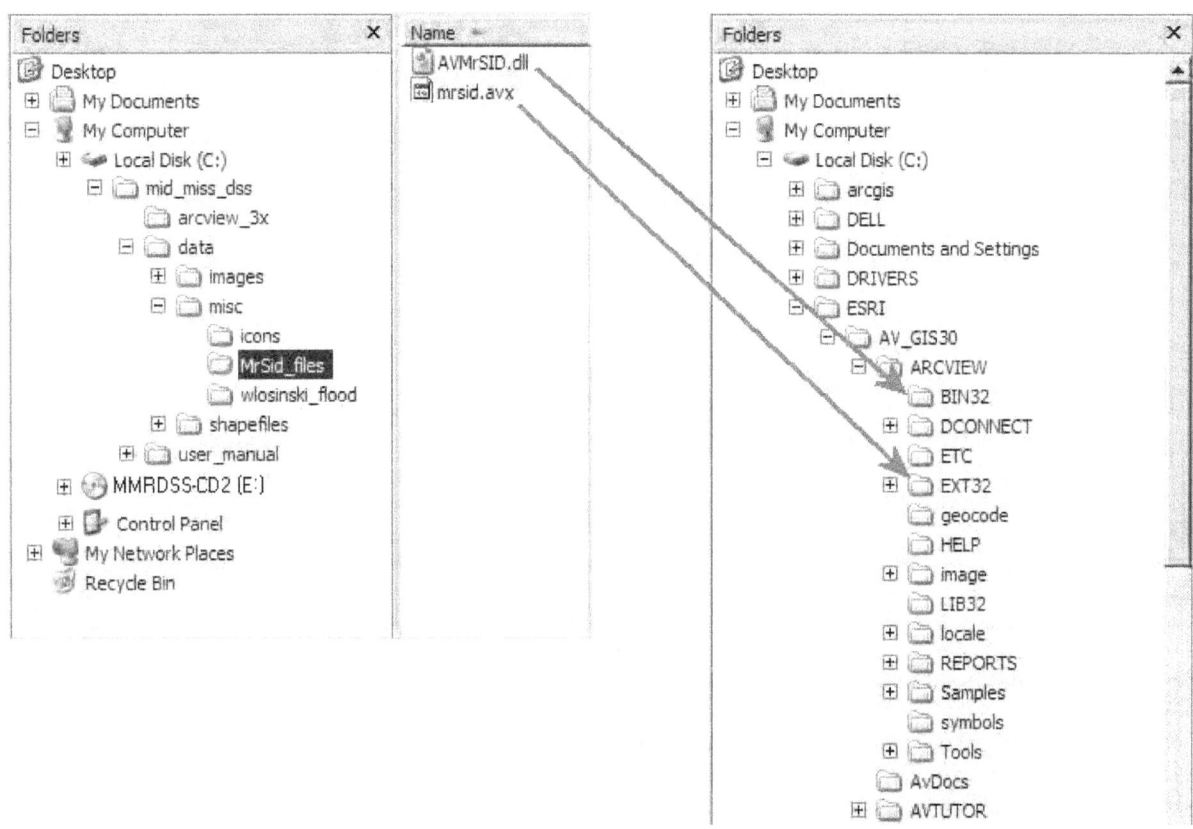

Figure 2–3. Copying the MrSid files to the hard drive.

Starting ArcView and Opening the MMRDSS

1. Click on the Windows **Start** button.
2. From the Start menu, navigate to **Programs -> ESRI -> ArcView GIS 3.x ->** and select "**ArcView GIS 3.x**" to bring up ArcView (fig. 2–4).
3. Once ArcView has started, select **Open Project...** from the **File** menu options (fig. 2–5).

Figure 2–4.   Starting ArcView.

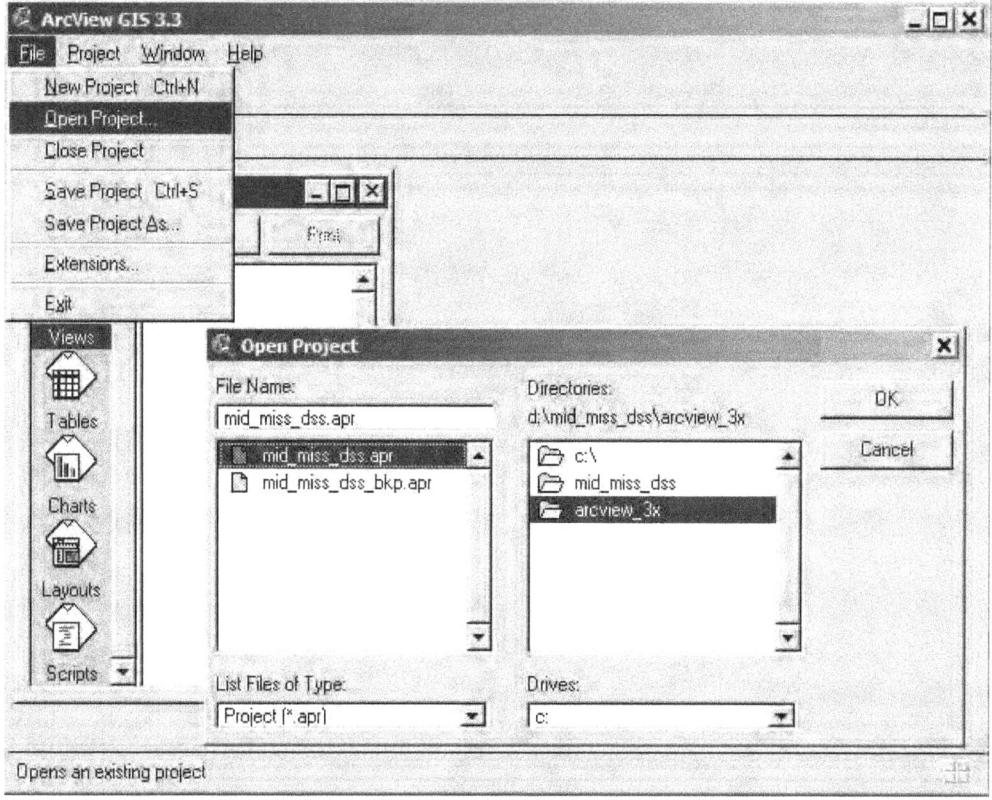

Figure 2–5.   Selecting Open Project.

4. In the Open Project dialog window, select ArcView project *\mid_miss_dss\arcview_3x\mid_miss_dss.apr* and click on the **OK** button (fig. 2–5).
5. The MMRDSS opens within ArcView with the MrSid Image Support extension activated.
6. There is a second ArcView project within *\mid_miss_dss\arcview_3x* called *mid_miss_dss_bkp.apr*. This is a backup copy of the original in case the original is deleted or corrupted.

Section 3: Data

GIS Data Themes

The GIS data (i.e., themes) were collected and added to the MMRDSS on the basis of their potential application to natural resource issues. The themes in the MMRDSS were compiled from private, state and Federal agencies, universities, and other individuals. **It is important to note that the themes contained within the MMRDSS vary in scale, coverage area, accuracy, and precision. The U.S. Geological Survey does not assume responsibility for inaccuracies with the themes collected. It is the user's responsibility to assess metadata and verify that the individual data themes are sufficient for their purposes.**

The format of each theme was checked to ensure seamless integration within the MMRDSS. This involved several steps.
1. Themes in formats different than an ArcView Shapefile were converted as follows:
    a. The ARC/INFO coverages were converted to shapefiles using ArcView 3.x.
    b. Intergraph design files from the U.S. Army Corps of Engineers were converted from their original format (.dgn) to shapefiles using ArcGIS ArcToolbox (ESRI; Redlands, CA).
    c. The ARC/INFO grids were converted to shapefiles using ArcView 3.x Spatial Analyst (ESRI; Redlands, CA).
    d. Point data in Excel spreadsheets were converted to shapefiles using ArcView 3.x.
    e. Large georeferenced image files were converted from their original format to Lizardtech's MrSID format (.sid), which compresses image files to minimize file storage space.
2. Themes in shapefile format were projected from their original spatial projection into the one selected for the MMRDSS using ArcView's Projector extension. This projection is described as follows:
    a. PROJECTION: Universal Transverse Mercator (UTM)
    b. UTM ZONE NUMBER: 15
    c. PROJECTION UNITS: Meters
    d. HORIZONTAL DATUM: North American Datum of 1927
    e. ELLIPSOID NAME: Clarke 1866

The projected themes were then added to the View document within ArcView. Appendix 2 contains a list of GIS data themes within the MMRDSS.

Metadata

Metadata were collected or created using ESRI's ArcCatalog for all the themes within the MMRDSS. These metadata were copied to the same directory on the hard drive as the shapefile or georeferenced image. The metadata for each theme can be viewed within the MMRDSS. To open the file for viewing, the desired data theme must be active within the MMRDSS View window. A theme is active if it is selected within the View's table of contents and a raised appearance surrounds the theme's name and legend (fig. 3–1).

To open the metadata file, select **Display Themes Metadata** from the **Mid_miss_tools** menu options. Alternatively, click on the **Display Metadata** button M . The metadata text file opens in a new ArcView window document (fig. 3–2).

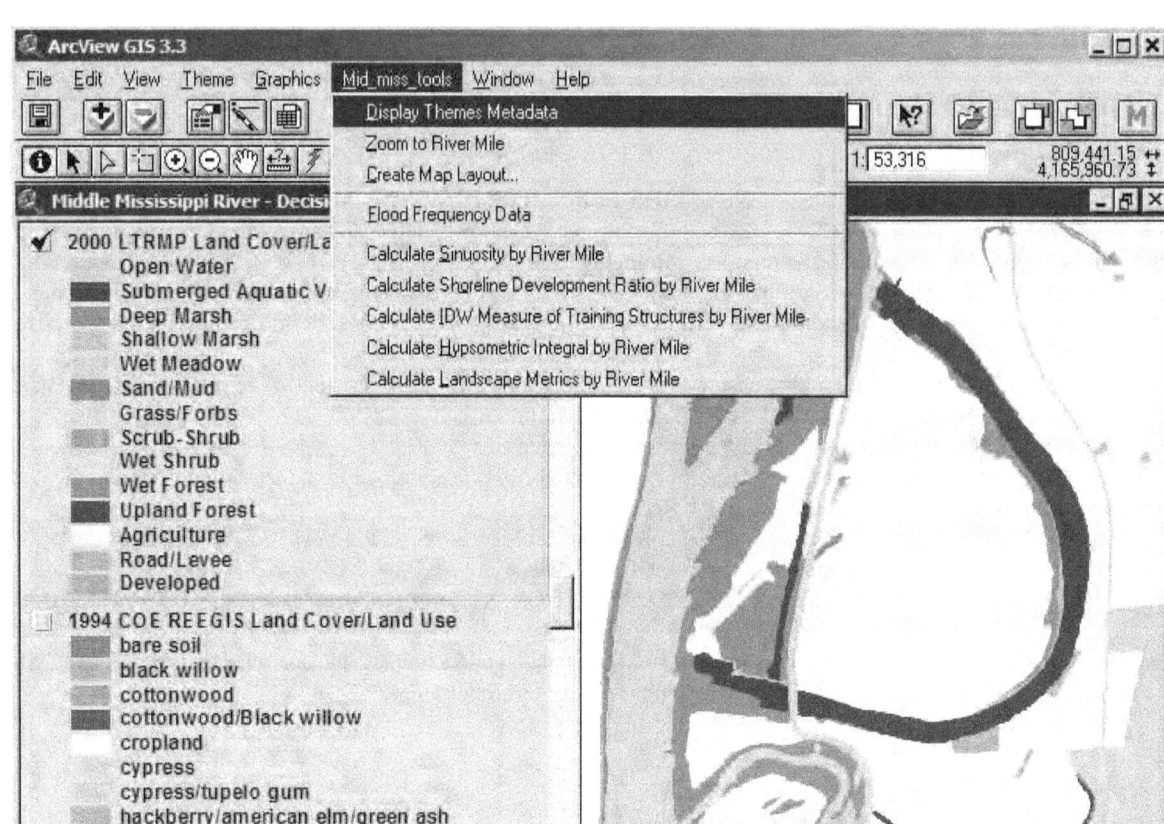

Figure 3–1.   Displaying the active theme's metadata.

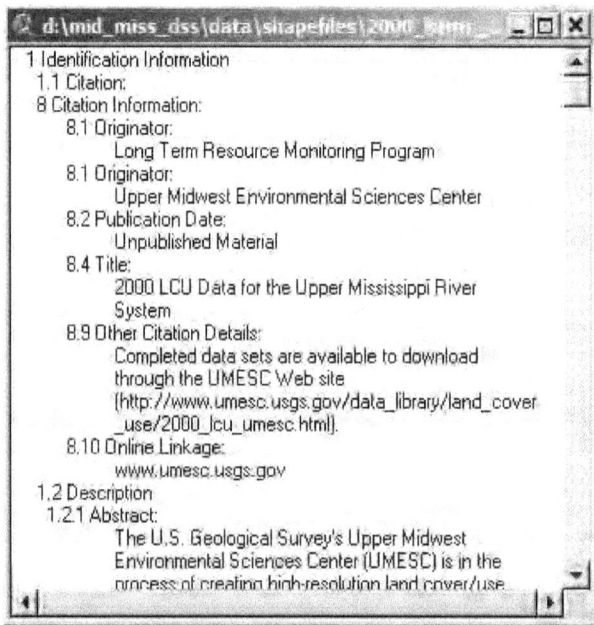

Figure 3–2.   Sample metadata file.

Section 4: Basic ArcView Functions

Creating Map Layouts

Map layouts for printing and viewing can be created using the MMRDSS. To create a map layout, make sure at least one theme is checked within the Vew document's table of contents.

1. Select **Create Map Layout...** from the **Mid_miss_tools** menu options (fig. 4–1).
2. Select the desired **map layout style** and click on the **OK** button (fig. 4–2).
3. A finished map layout opens with the themes visible within the view added along with a legend, title, north arrow, and scale bar (fig. 4–3). Elements within the layout (map, title, legend, scale bar, north arrow) can be moved and resized by selecting them with the **Pointer Tool** .

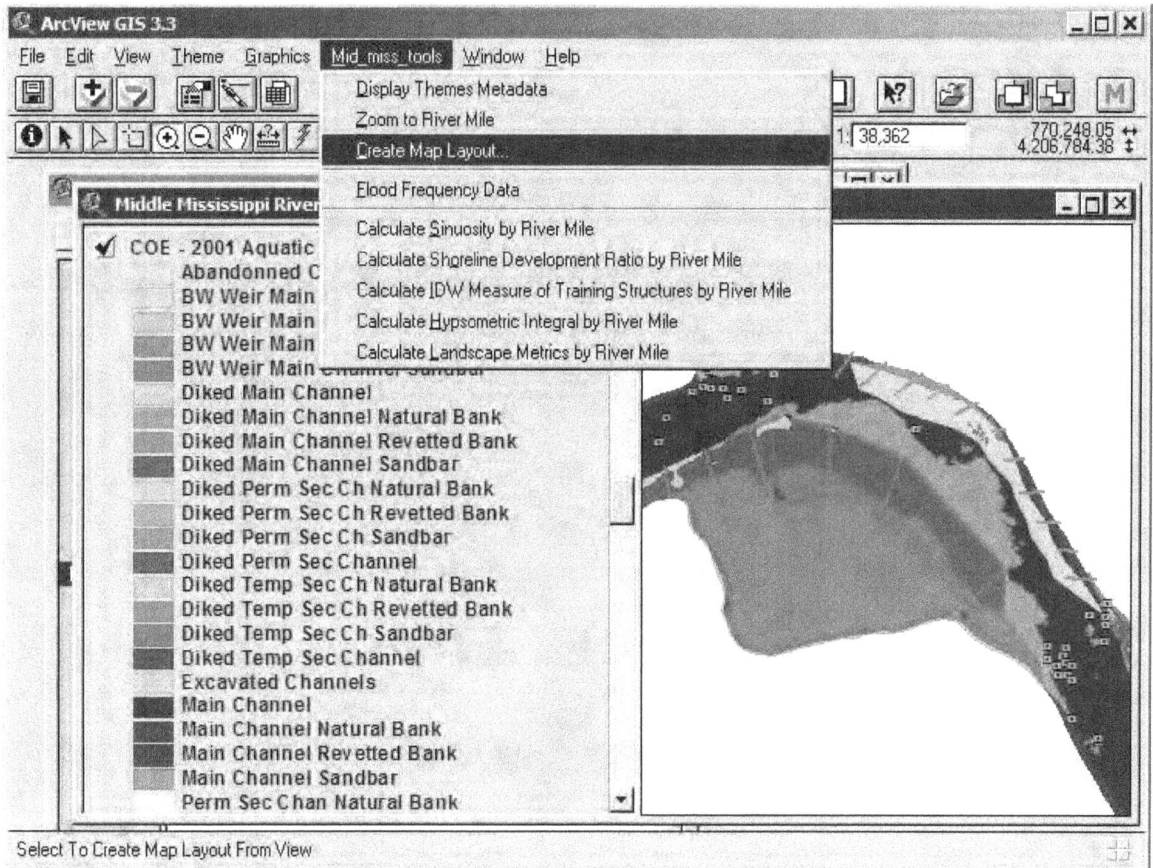

Figure 4–1. Creating a map layout.

Copying Graphics to Other Windows-Based Programs

Layout graphics generated by the MMRDSS can be copied to other computer programs, such as Microsoft PowerPoint.

1. Click on the **Pointer Tool** .
3. Select **Copy** from the **Edit** menu options.
4. Start Microsoft PowerPoint and insert a new slide if one does not exist.
5. Select **Paste** from the **Edit** menu options. The copied graphics are now displayed on the PowerPoint slide. The graphics can be moved and resized as necessary.

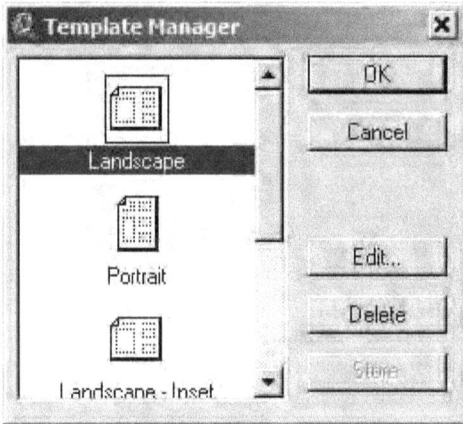

Figure 4–2.   Select a layout format from the Template Manager.

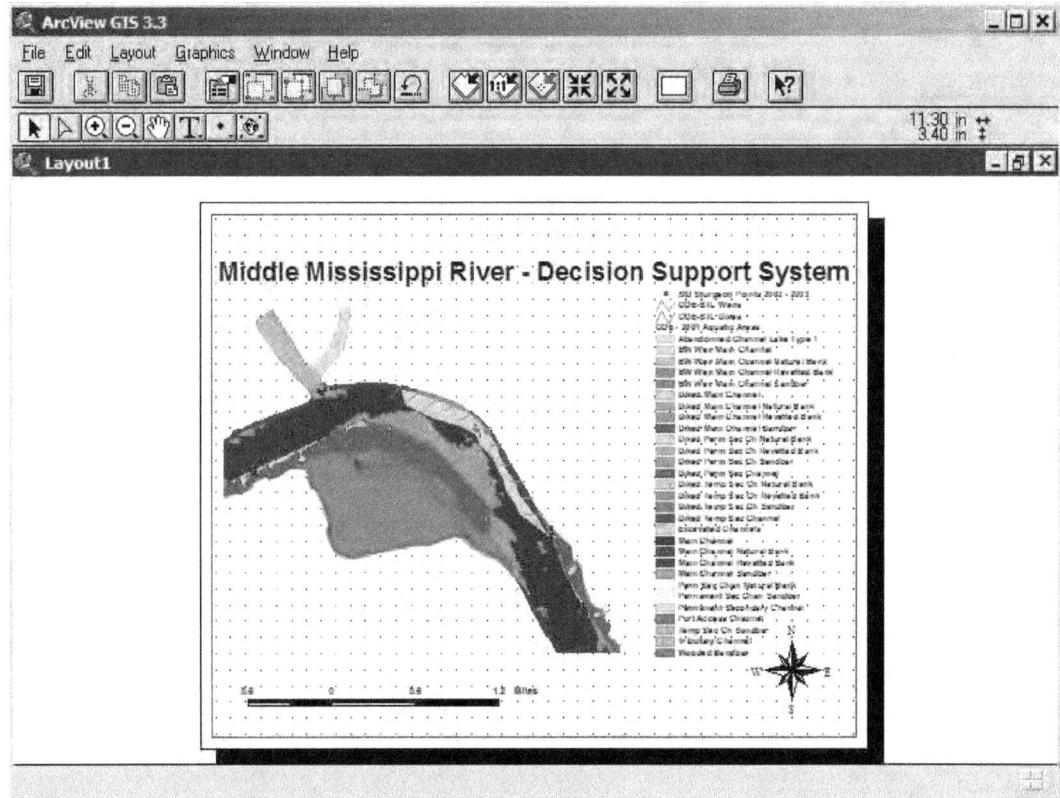

Figure 4–3.   Finished map layout.

6.   The entire map display can also be exported from ArcView as a digital image file (e.g., Windows Bitmap [.bmp], JPEG [jpg], etc.) by selecting **Export…** from the **File** menu options.
7.   A window opens prompting the user to select a **filename**, **destination**, and **file type**.

Exporting Tables to Other Windows-Based Programs

Tables within ArcView can be exported to other programs, such as Microsoft Excel, for example:

1.   Within the ArcView Project Window, click on the **Tables** icon (fig. 4–5). Select the *UMESC-LTRMP Fall 1997 Fisheries Query Table* and click on the **Open** button.

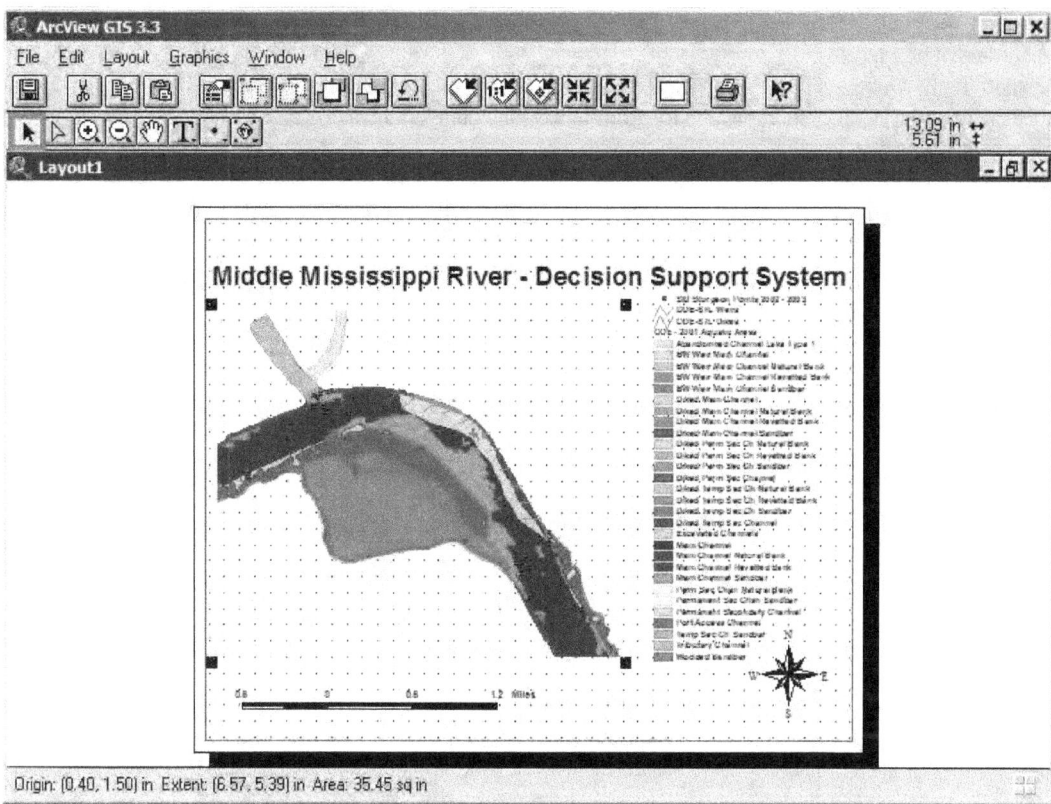

Figure 4–4.   Selecting graphics to copy from the map layout.

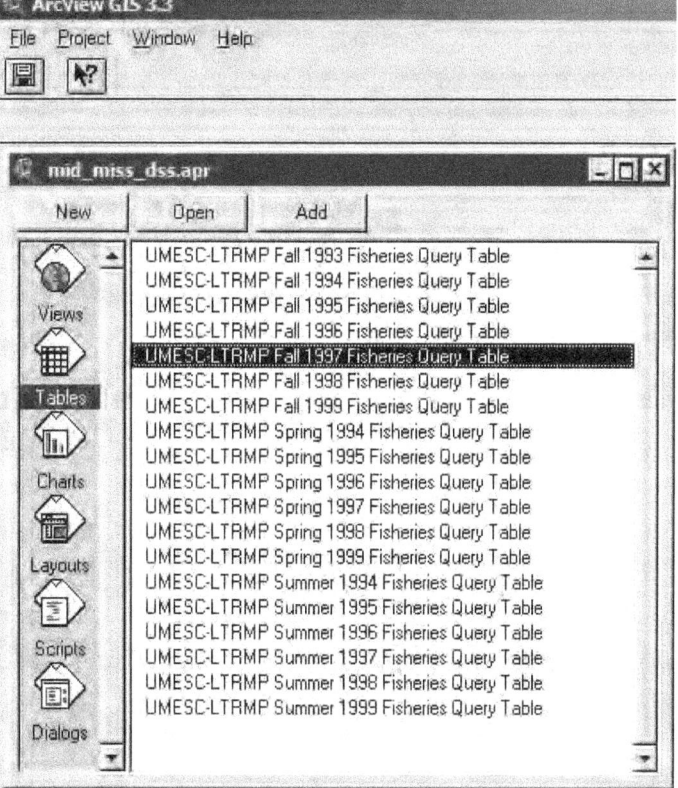

Figure 4–5.   ArcView's project window.

2. With the table open, select **Export...** from the **File** menu options (fig. 4–6). If any records are selected within the table, only those records will be exported.
3. In the Export Table window, select the **Delimited Text** format and click on the **OK** button.
4. In the next window, select the **filename** and **location** to save the table on the hard drive.
5. Start Microsoft Excel.
6. In Excel, select **Open** from the **File** menu options and set the ***Files of Type:*** box to **All Files** (*.*).
7. Navigate to the directory where you saved the file in step 4, select the **exported table file** (.txt), and then click on the **Open** button.
8. Once the Text Import Wizard window opens, choose **Delimited** as the original data type and click on the **Next** button.
9. Click the check box next to **Comma** and click on the **Finish** button. The table is now loaded into Microsoft Excel as a spreadsheet.

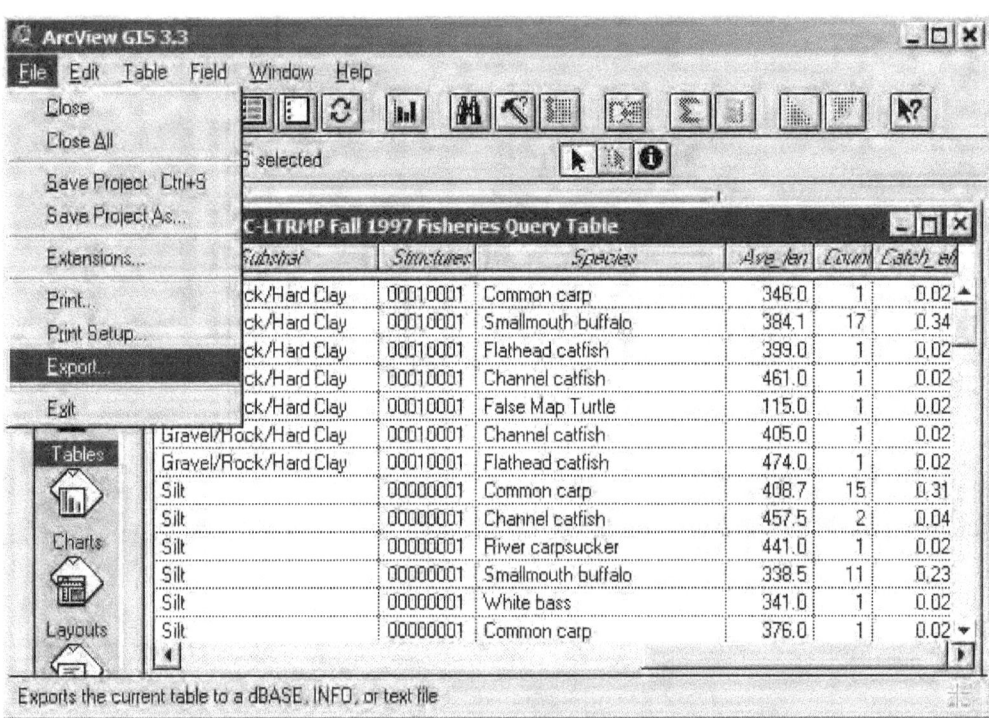

Figure 4–6. Exporting tables.

Using Standard ArcView GIS Buttons and Tools

ArcView GIS has several standard buttons and tools that may be used to alter the display, query the data, and in other ways enhance the capabilities of the MMRDSS. These buttons and tools are accessed using ArcView's graphical user interface (fig. 4–7). Make the theme *2000 LTRMP Land Cover/Land Use* active and visible (checked) for the following exercises.

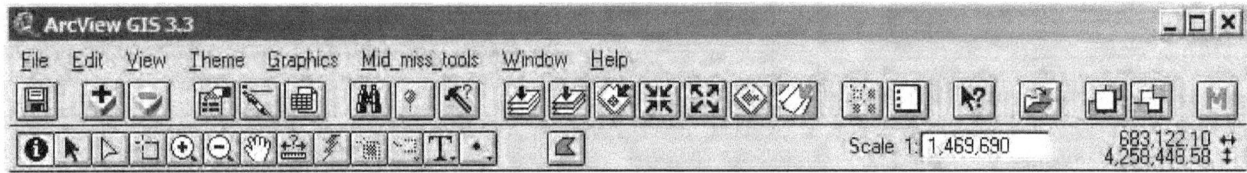

Figure 4–7. ArcView Graphical User Interface for View documents.

The **Zoom In** [⊕], **Zoom Out** [⊖], and **Pan** [✋] **Tools** (fig. 4–8) allow the user to change the display within the View window. The **Zoom to Active Themes** button [⬧] changes the display to the geographic extent of the active theme(s) within the View.

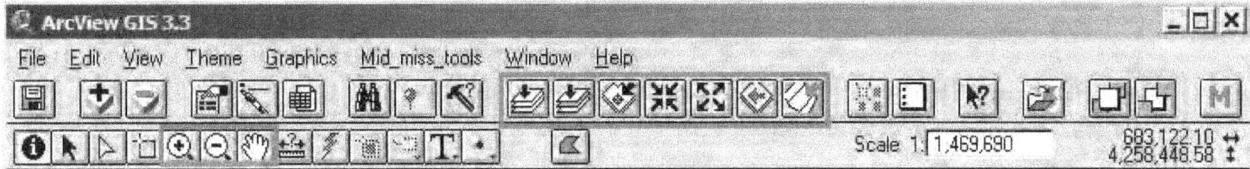

Figure 4–8.   Zoom In, Zoom Out, Pan, and Zoom to Active Themes tools and button.

By selecting **Zoom to River Mile** from the **Mid_miss_tools** menu options, the display within the view can be changed to view only the extent of the two river miles selected (fig. 4–9). This function can also be accessed by pressing the **Zoom to River Mile Tool** [⬧].

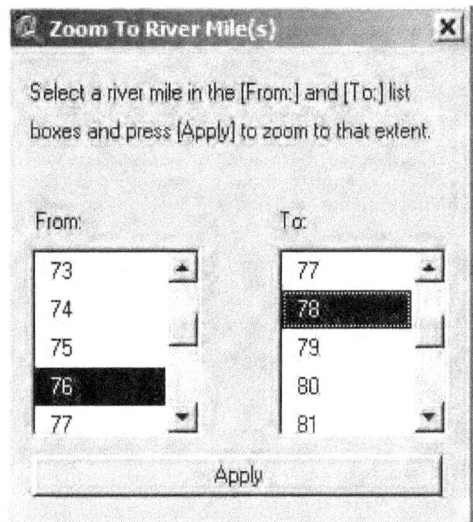

Figure 4–9.   Zoom to River Mile(s) window.

The **Identify Tool** ⓘ (fig. 4–10) displays the tabular information about a theme's feature. This tool is used by clicking an **active themes feature** with the tool depressed.

The **Label Tool** (fig. 4–11) draws textual tabular information on the view. To designate the table attribute labeled within the display, select **Properties** from the **Theme** menu options. Click on the **Text Labels** icon and select the **field** from the drop-down menu for the *Label Field* (fig. 4–12).

The **Measure** and **Draw Tools** (fig. 4–13) allow linear and area measurements within the view. For this example, the **Draw Polygon Tool** was selected.

The **Query Builder** selects records within a theme's feature attribute table, and these selected records are displayed as highlighted records within the table and highlighted features within the view. Make the theme *UMESC-LTRMP Summer 1994 Fisheries Component Data* active and visible by checking it within the View's table of contents. Next, click on the **Query Builder** button [⬧]. We included additional programming for the MMRDSS to prompt the user to either **Query Linked Query Table** or to **Query Feature Attribute Table** for the *UMESC-LTRMP Fisheries Component Data* themes. This was done because there are several records for each sampling point. The collected sampling data cannot be included within the theme's feature attribute table that represents a one-to-one relation. Conversely, the linked table can support many-to-one relationships. For this exercise, select the option **Query Linked Query Table** (fig. 4–14).

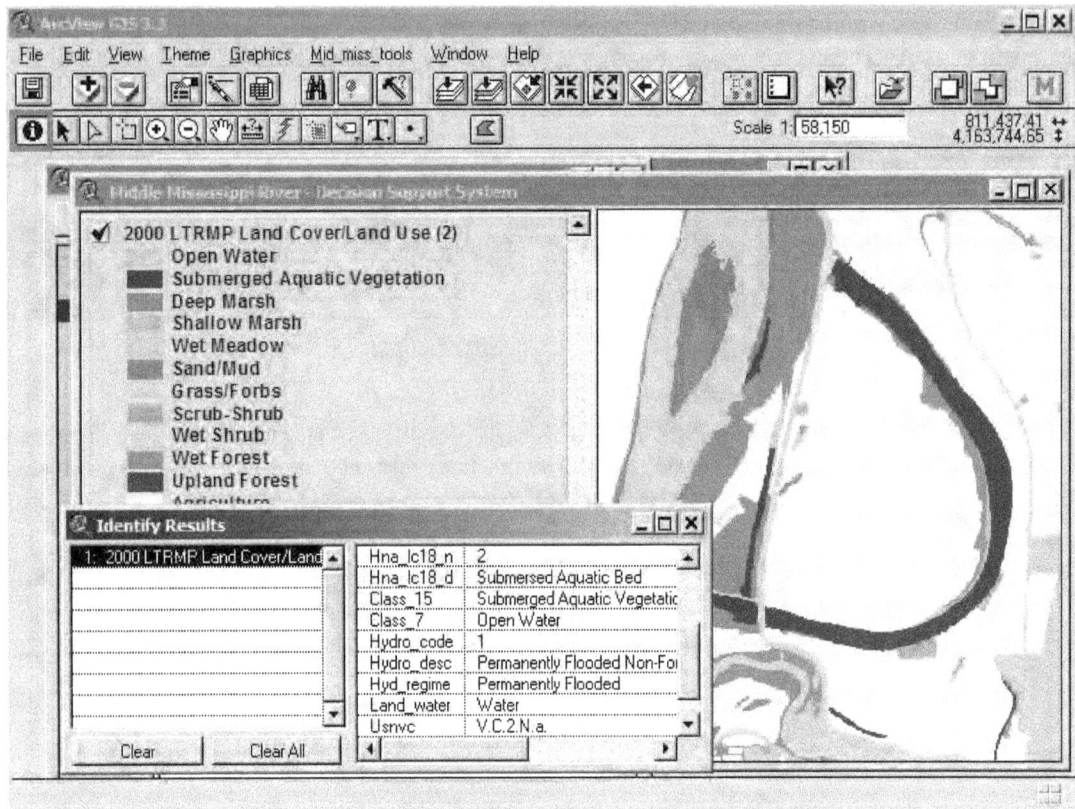

Figure 4–10. Identify tool.

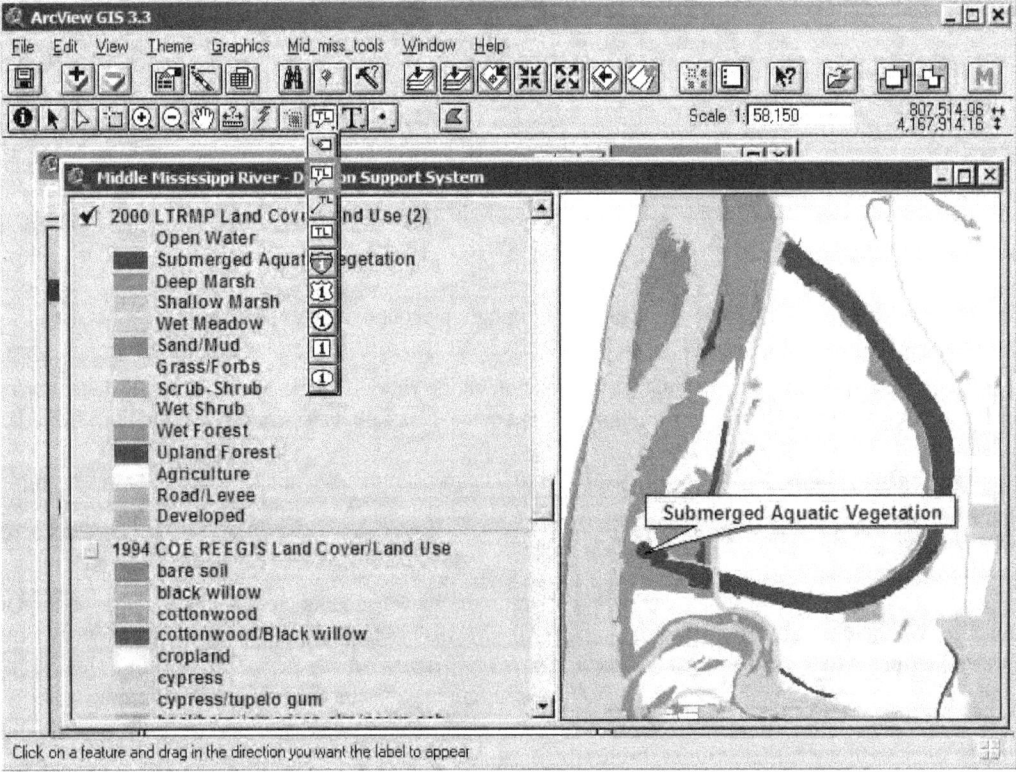

Figure 4–11. Label tool.

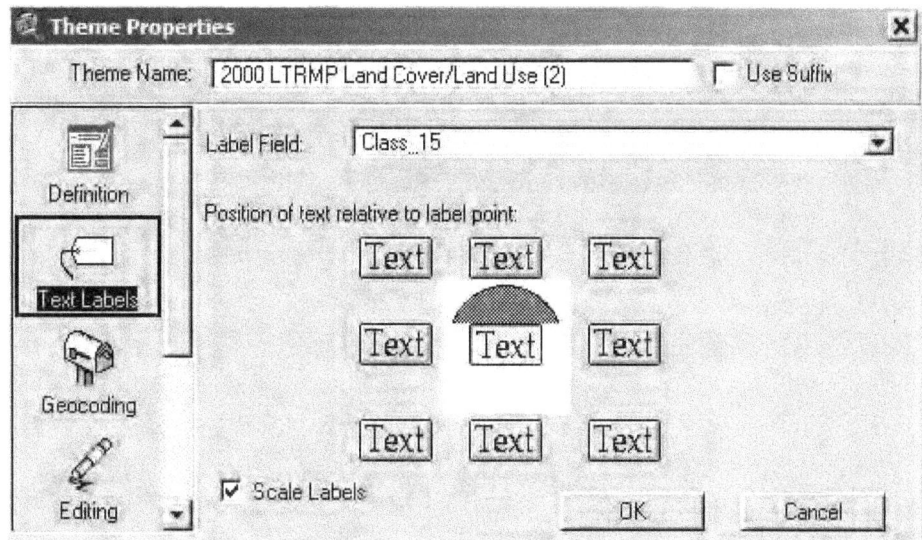

Figure 4–12. Setting theme properties.

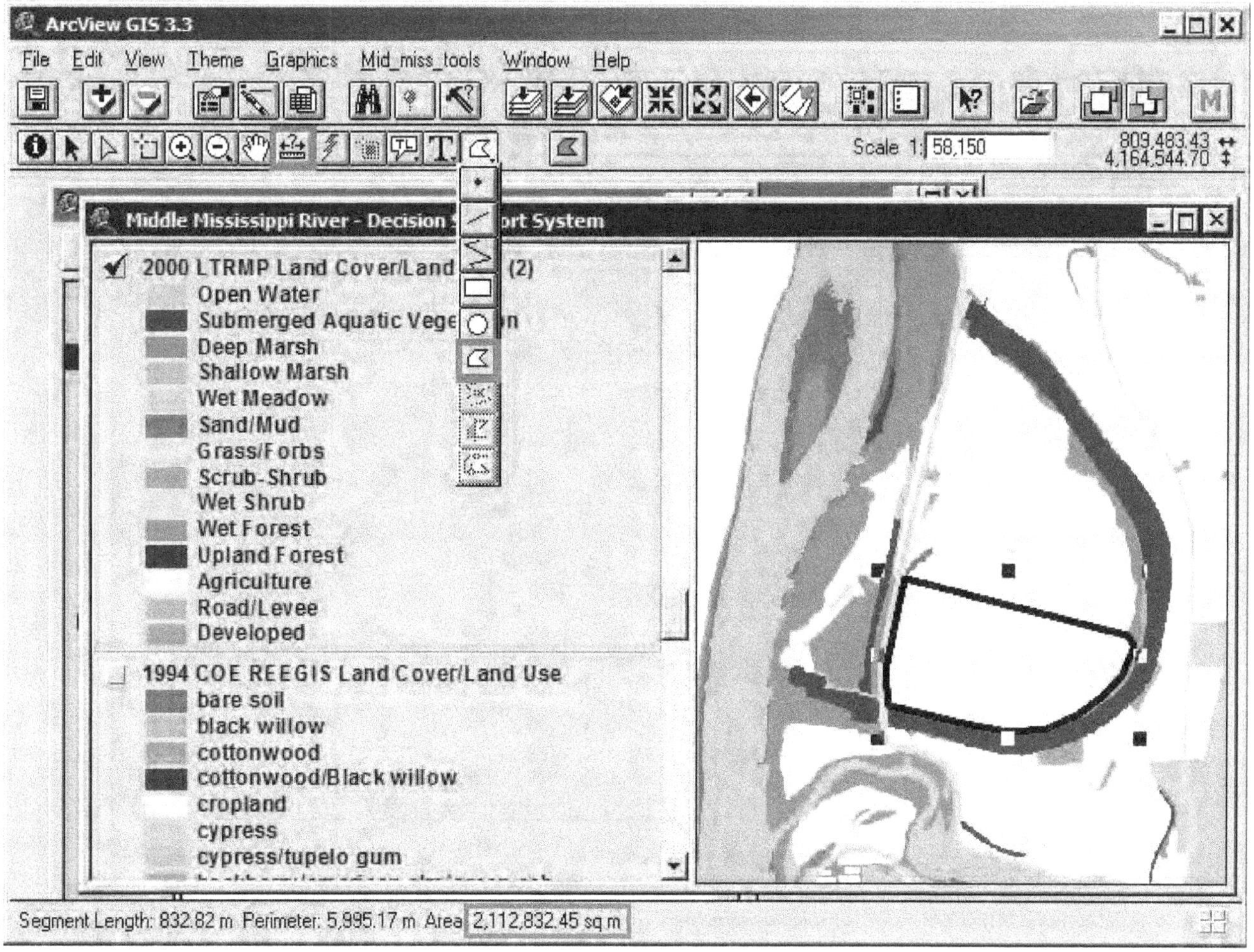

Figure 4–13. Measure and Draw tools.

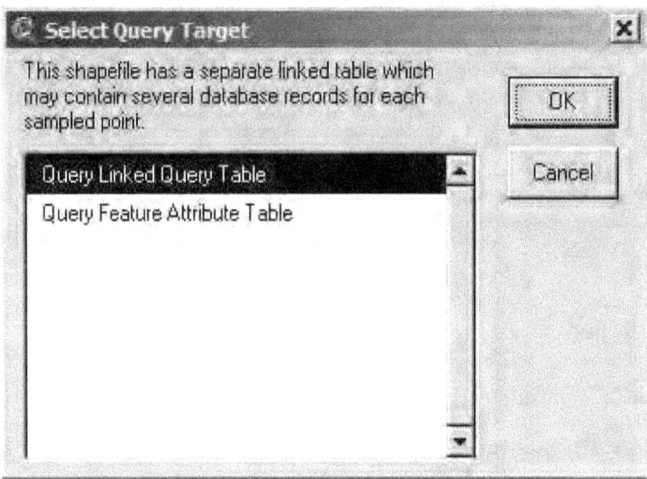

Figure 4–14. Select Query Target.

Within the Query Builder window, construct the query (*[Species] = "Shovelnose sturgeon"*) and click on the **New Set** button (fig. 4–15). Figure 4–16 displays the results of the query.

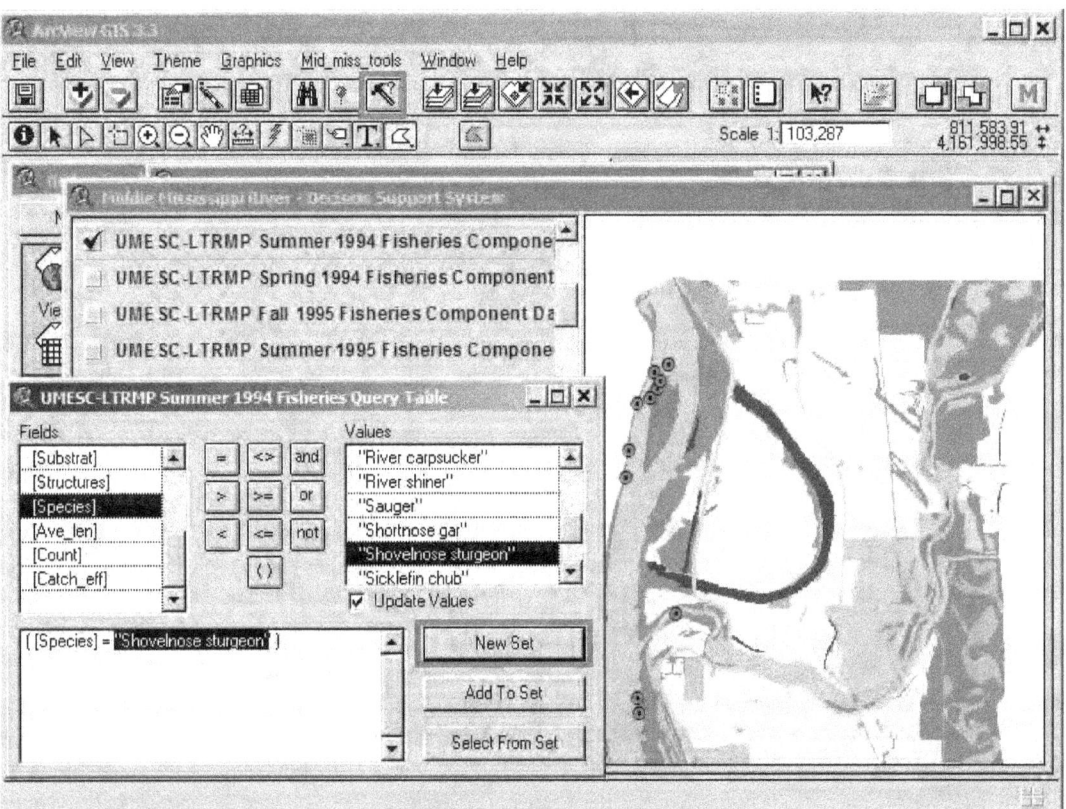

Figure 4–15. Query Builder.

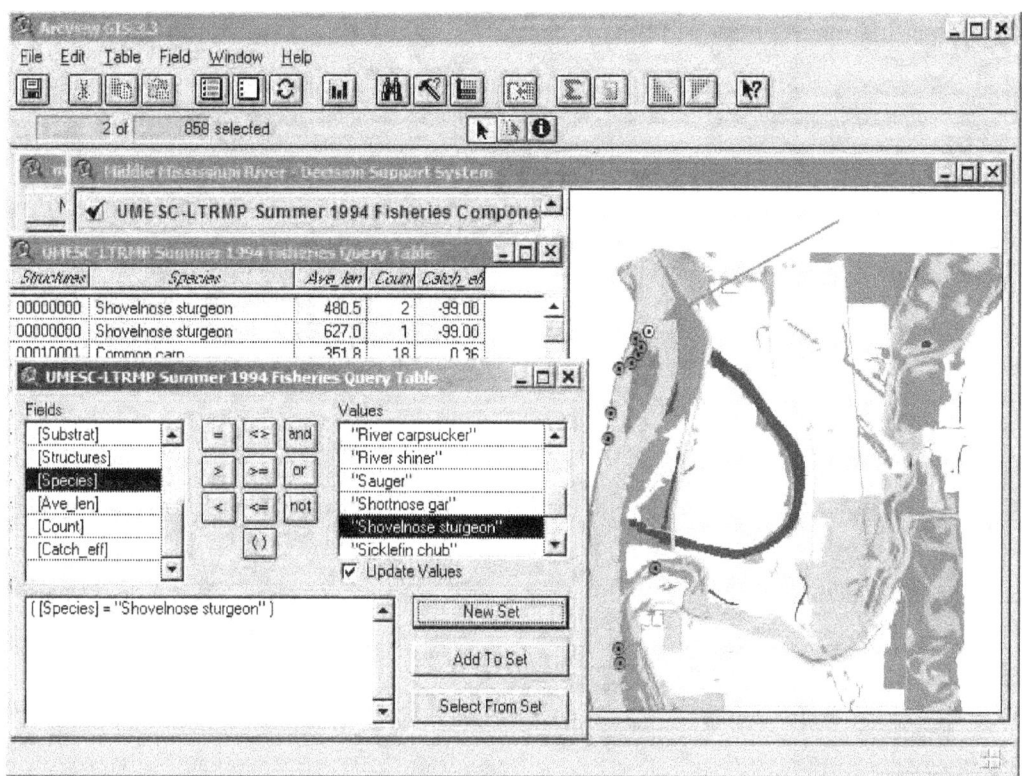

Figure 4–16.   Query Builder Results.

Other GIS Data Theme Functions

The GIS Data Themes, besides those existing within the View, can be added using the **Add Theme** button ![icon].

Themes that already are loaded within the View can also be deleted by selecting those themes and pressing the **Delete Theme(s)** button ![icon], a programming addition to the MMRDSS.

Themes within the View's table of contents can be moved from any position to either the top of the table of contents ![icon] (fig. 4–17) or to the bottom of the table of contents ![icon]. If multiple themes are selected, they all will be moved.

Figure 4–17.   Active Theme(s) to Top.

Section 5: Advanced Tools–Geomorphology and Aquatic Habitat

The MMRDSS **Mid_miss_tools** menu options (fig. 5–1) provide access to a suite of program tools to assist users in evaluating differences in complexity, connectivity, and structure of aquatic habitats among river reaches. These program tools, which were developed for the MMRDSS as ArcView Avenue scripts, generate information and GIS themes derived from floodplain planforms, including flood frequency, sinuosity, hypsometric integral, and landscape metrics. Justification and proposed hypotheses for development of tools (except flood frequency data) are provided in Hulse and Galat (2003). Tool descriptions include excerpts from that document.

The biological relevance of the metrics calculated by these tools has not been tested in the Middle Mississippi River, but is grounded in sound ecological theory and evidence that floodplain planform influences biotic communities and processes in rivers (Johnson and others, 1995; Amoros and Bornette, 2002; Fausch and others, 2002; Wildhaber and others, 2003; Koel, 2004). Because biological and physical processes in rivers are strongly tied to spatial scale and the metrics are scale-dependent, the tools incorporate flexibility to calculate metrics at various scales. It is up to the user to select the appropriate scale to address management and research questions.

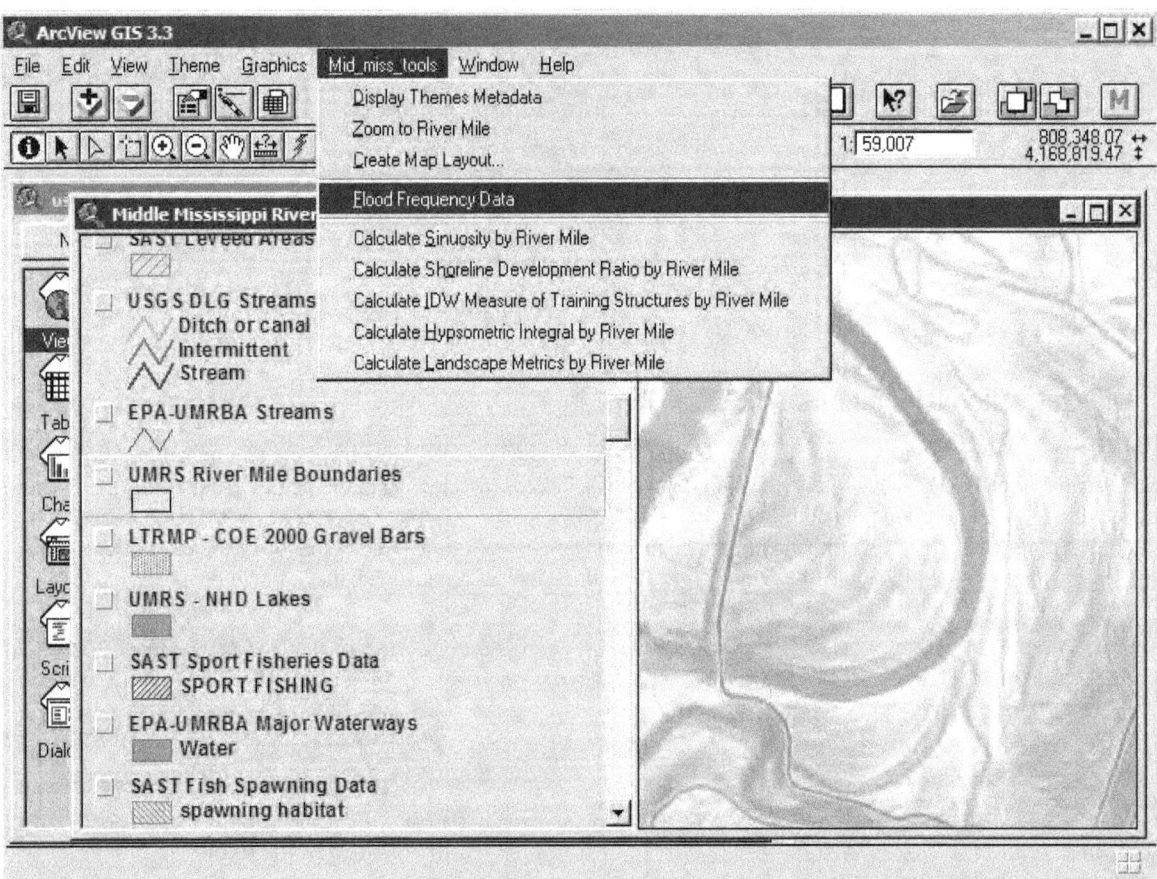

Figure 5–1. Wlosinski and Wlosinski (2001) Flood Data menu item.

Flood Frequency Data

Tabular data included within the MMRDSS displays the percentage of chance for flooding per year, at different elevations, and for different durations within the Middle Mississippi River (Wlosinski and Wlosinski, 2001).

1. To display these data, select **Flood Frequency Data** from the **Mid_miss_tools** menu options (fig. 5–1).
2. Next, **Select a River mile to view the corresponding flood data** (fig. 5–2) and click on the **OK** button.

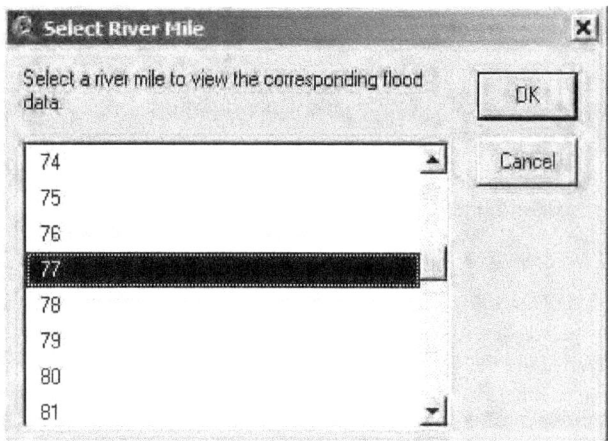

Figure 5–2. Select a River Mile.

3. A table opens displaying the elevations (in feet above mean sea level) and the percent chance of flooding for different durations at that elevation. A text file also opens (fig. 5–3) with background information and methods used to calculate flood frequencies (Wlosinski and Wlosinski, 2001).

| % chance of flooding at different elevations and for different durations for river mile 77 | | | | | | |
Elevation	3 days	1 week	2 weeks	4 weeks	6 weeks	8 weeks
370	2	0	0	0	0	0
369	2	2	0	0	0	0
368	2	2	2	0	0	0
367	2	2	2	0	0	0
366	4	2	2	2	0	0
365	7	4	2	2	0	0
364	7	7	4	2	0	0
363	9	7	4	2	2	0
362	11	7	7	2	2	0
361	13	9	7	4	2	2
360	16	13	9	4	2	2
359	20	16	9	7	2	2
358	22	16	11	7	4	2
357	29	22	16	7	4	2
356	31	24	18	11	4	2

Background Information

```
Percentage chance of flooding per year, Upper
Mississippi River, at different elevations and for
different durations.  For each duration, percentage
estimates may include multiple flooding events
within a single year.  Elevations are reported as
feet above sea level (1929 National Geodetic Vertical
Datum).  Analyses were based on U.S. Army Corps of
Engineers water surface elevation data collected from
1954 to 1998 (inclusive), for the estimated growing
season between April 1 to November 1 and were conducted
by the Upper Midwest Environmental Sciences Center and
the Upper Mississippi River National Wildlife and Fish Refuge.

The project status report (PSR) for this study is
located in the directory
\mid_miss_dss\data\misc\wlosinski_flood\psr01_01.pdf
and can also be downloaded at the website
http://www.umesc.usgs.gov/data_library/water_elevation/flood_potential.html
```

Figure 5–3. Flood Data Output.

Sinuosity Index

River channel sinuosity may characterize channel planform complexity when evaluated at an appropriate spatial scale. This tool calculates the Sinuosity Index (SI) of a given river reach using two themes. These themes do not need to be added to the view.

- The first theme is a polyline representing the channel thalweg. For the MMRDSS, the U.S. Army Corps of Engineers' sail line was used to approximate the Middle Mississippi River thalweg.

- The second theme consists of focal points along the thalweg where SI will be calculated.

The equation for SI used in the MMRDSS is based on Leopold and others (1964), where SI is calculated as the ratio of channel length to down valley length (SI = L_s/L_v; fig. 5–4). The tool calculates SI at each focal point by measuring the user-specified distance upstream and downstream along the channel (L_s; meandering channel length) then dividing by the linear distance between these points not following the channel (L_v; down valley length).

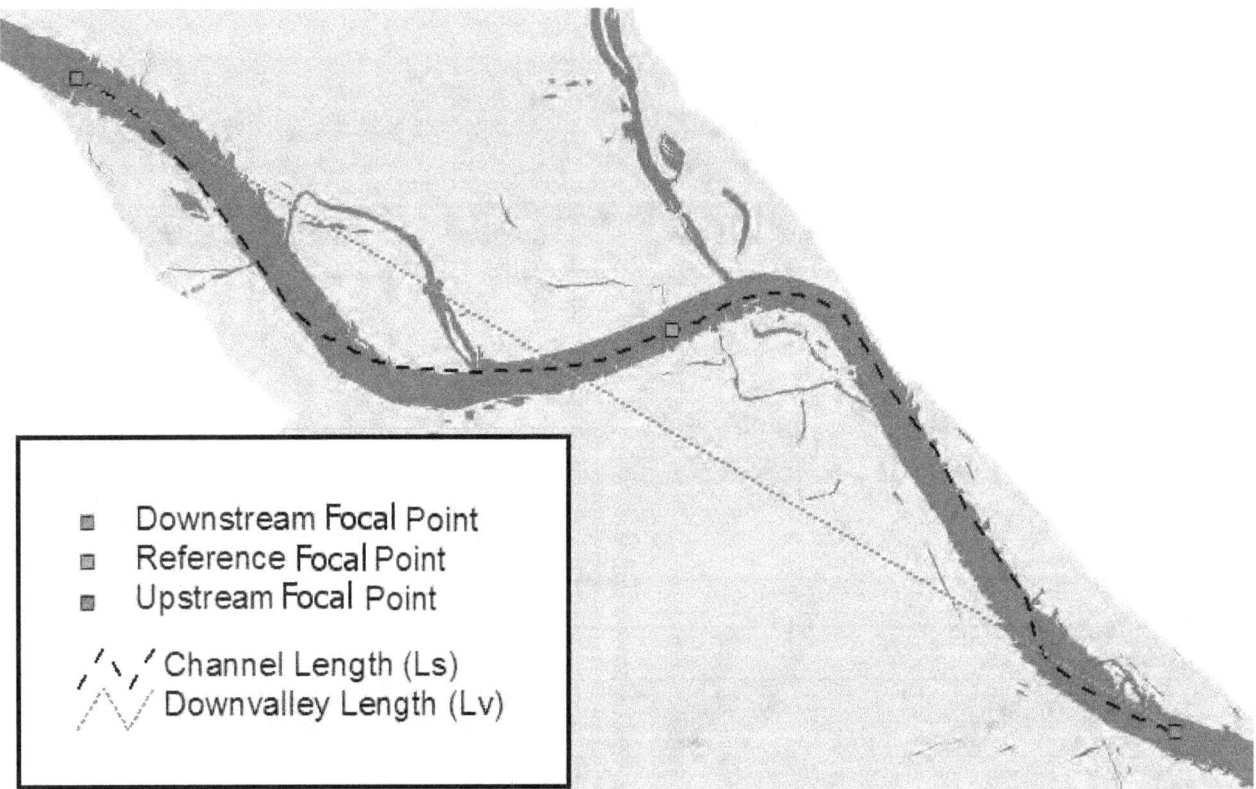

Figure 5–4. Calculating Sinuosity Index.

Because scale is an important factor in calculating SI, the **Sinuosity Index Tool** provides users with four scale options. As an extreme example, consider the channel length to be near zero. In this scenario, the index is not useful as an explanatory variable of planform complexity because its expected value is near 1.0 ($L_s \sim L_v$). Conversely, the single value of SI characterizing an entire river is expected to be ≥1.0, which is only useful when comparing variability among rivers.

1. To start the tool, select **Calculate Sinuosity by River Mile** from the **Mid_miss_tools** menu options (fig. 5–5).
2. Next, **Select total distance upstream and downstream of each focal point along river thalweg (scale)** for calculating the Sinuosity Index (fig. 5–6). The first three options have already been preprocessed for the user using the **Sinuosity Index Tool**. This was done so the user could bypass lengthy processing times and rapidly access these results. The distance between focal points for these scale options was 100 meters.

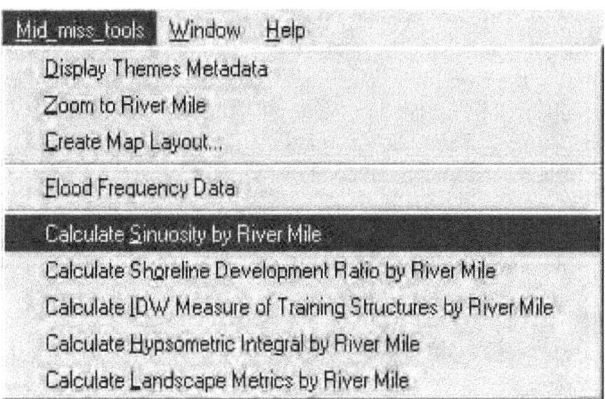

Figure 5–5.   Starting Sinuosity Index Tool.

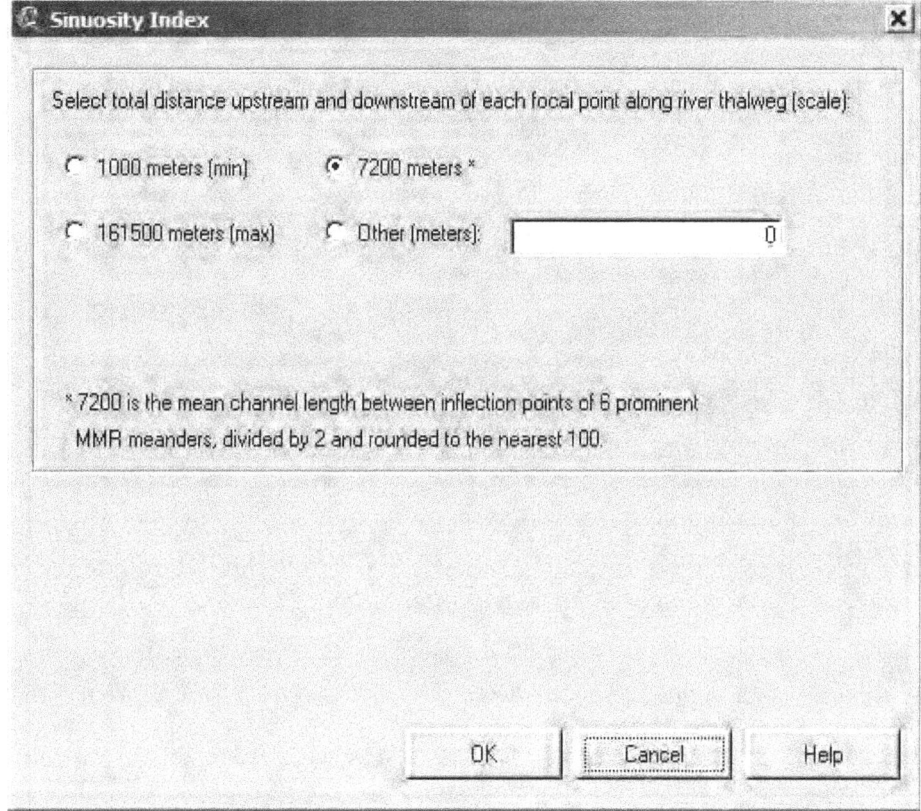

Figure 5–6.   Sinuosity Index tool dialog.

a.   The first option is **1000 meters (min)**, which was determined to be the minimum useful value for the scale of the Middle Mississippi River. However, values below 1000 meters may be appropriate for other highly sinuous streams.

b.   The second option is **7200 meters ***. This value represents the mean channel length between inflection points of six prominent Middle Mississippi River meanders, divided by two, and rounded to the nearest 100.

c.   The third option is **161500 meters (max)**. This is the maximum value and represents the total length of the Middle Mississippi River, divided by two. Selecting this option produces only one point and one polygon within the output themes. This number represents the SI of the Middle Mississippi River as a whole.

d. The last option is ***Other (meters):***. This option allows the user to input their own scale into the equation and also to choose from three different focal point distances (100, 200, and 400 meters). **Selecting this option requires substantially more processing time by the computer**. The user must select an integer value between 1000 (min) and 161500 (max) for scale. The scale value will be rounded to the nearest 100, 200, or 400 meters depending on the selected focal point distance that was selected. Figure 5–7 displays the Sinuosity Index Tool's input dialog window when the ***Other (meters):*** option is selected; notice the focal point distance options are now visible.

3. Click on the **Help** button to display a text file describing the use of the SI Tool (fig. 5–8).
4. Select the scale option ***Other (meters):*** and enter **10000** in the text box to the right.
5. Set the distance between each focal point to be **400 meters** and click on the **OK** button (fig. 5–7).

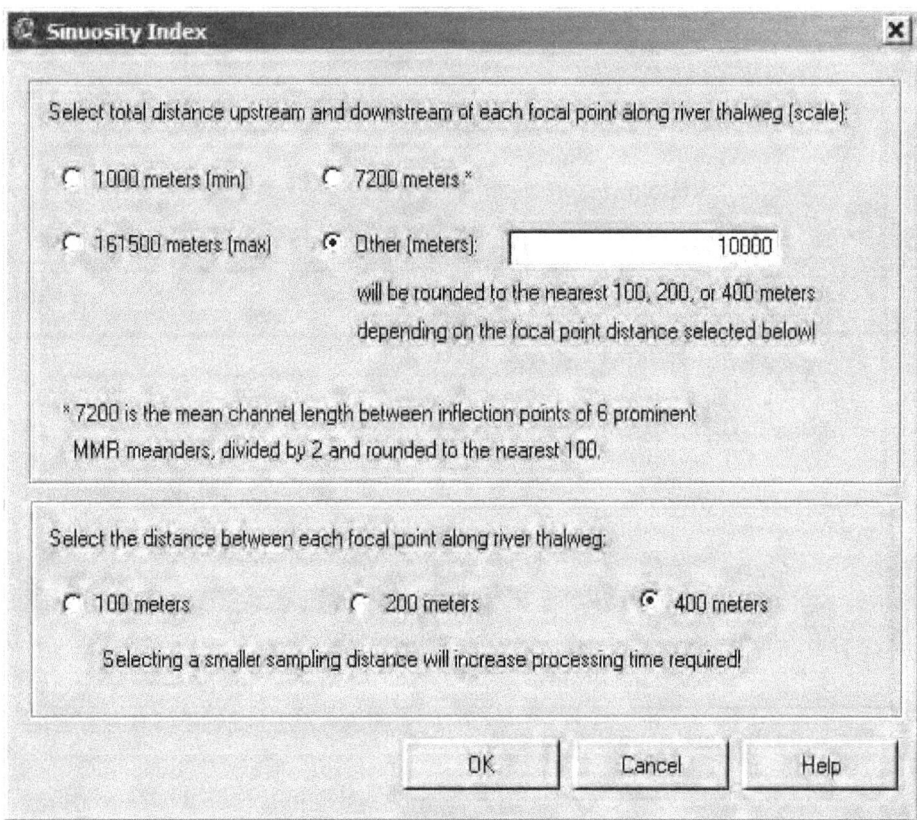

Figure 5–7. Sinuosity Index tool dialog displaying focal point distance options.

6. A window opens prompting where on the computer's hard drive to save the outputs. Select a **directory** and click on the **OK** button (fig. 5–9).

Once ArcView has processed the scripts, the output data are saved to the specified directory and two output shapefiles are added to the view (fig. 5–10).

This tool creates a theme, *Sinuosity Index Points (10000m)*, representing focal points spaced 400 meters apart along the sail line. This theme is attributed with the SI calculated at the scale selected (*Si*), the mean SI using three spatial scales (*Si_3scales*), and the total distance traveled upstream to each individual focal point (*Walkdist*).

To incorporate scale-dependent variability, the mean SI was calculated for three spatial scales including the entire Middle Mississippi River (RM 1–202 and SI=1.4099), the Long Term Resource Monitoring Program (LTRMP) Open River Reach (RM 29–80 and SI=1.2245), and the SI at each point along the sail line using mean meander lengths of (14440 m) in the numerator.

The larger the SI, the more sinuous that particular river segment is. The tool also creates a polygon theme *Sinuosity Index by RM (10000 m)* representing river mile segments with each separate polygon attributed by the average sinuosity index (*Si_ave*) of all the focal points that fall within that particular river mile polygon, the mean SI using three spatial scales (*Si_3scales*) for the entire river mile polygon, and the river mile number for each polygon (*River_mile*). Metadata are created for each theme and are accessible by clicking on the **Display Metadata** button [M].

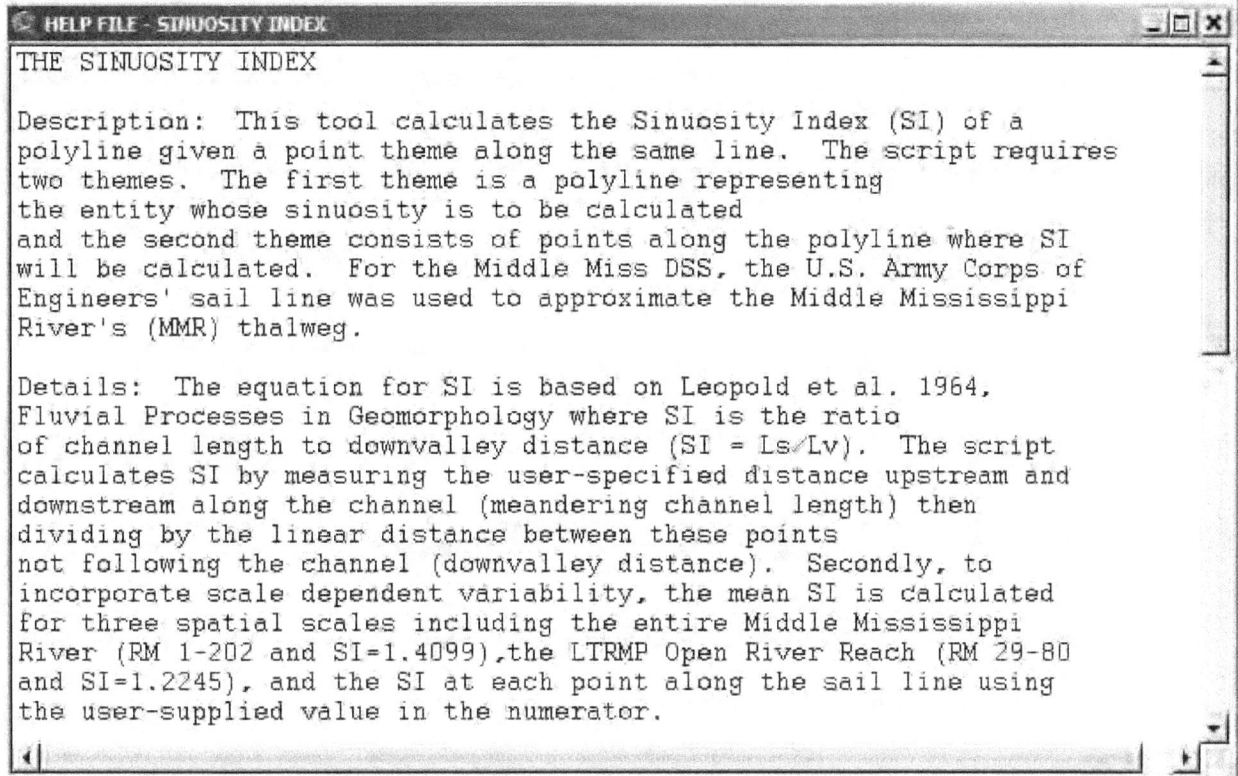

Figure 5–8.   Sinuosity Index Help File.

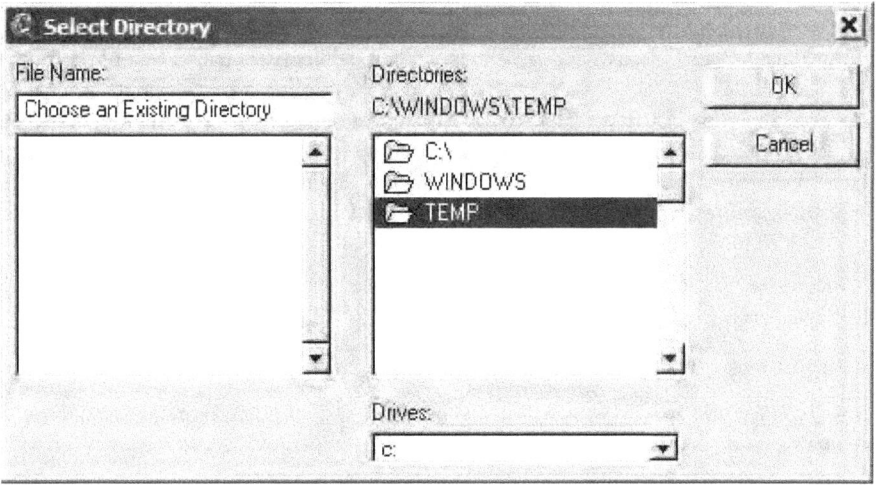

Figure 5–9.   Specify Output Directory.

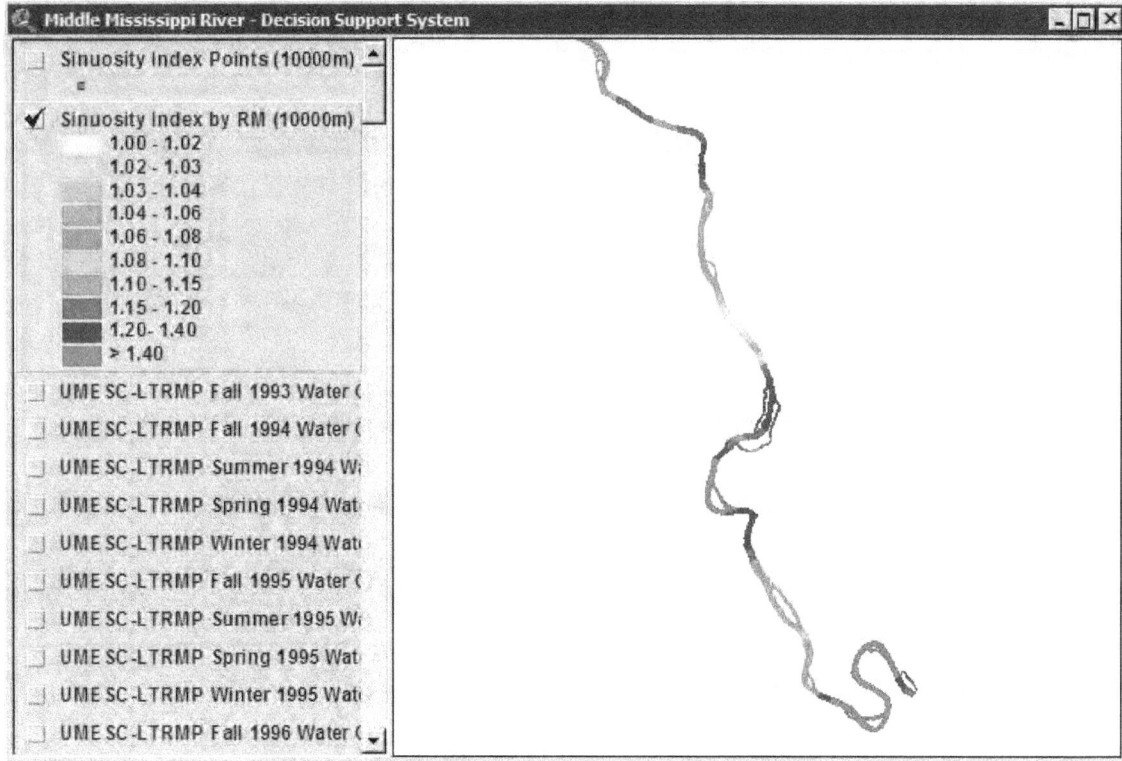

Figure 5–10.   Sinuosity Index Outputs.

Shoreline Development Ratio

Shoreline convolution may alter physical and biological processes within a water body and indicate heterogeneity and complexity of aquatic habitats. Thus, shoreline convolution may correlate with species richness or abundance. A water body characterized by a highly convoluted shoreline may have greater richness of habitat types than a water body with a nonconvoluted shoreline.

This tool calculates the Shoreline Development Ratio (SDR) using a polyline theme to approximate the study area shoreline. Three themes are used to calculate SDR within the tool. The first theme is a reference polyline that the program travels upstream as it calculates SDR. The second theme consists of focal points along the polyline where each SDR is calculated. The third theme is a polyline theme representing the shoreline. For the MMRDSS, three themes were used to determine SDR. These themes do not need to be added to the view.

- For the first theme, the U.S. Army Corps of Engineers' sail line was used to approximate the Middle Mississippi River's (MMR) thalweg.

- The second theme was constructed for the MMRDSS using a script that places points evenly along the sail line.

- The third theme is a polyline theme developed using the 2000 Upper Midwest Environmental Sciences Center (UMESC)/LTRMP Land Use/Land Cover representing the shoreline of the Middle Mississippi River.

The SDR is calculated similar to shoreline development for lakes as described by Forman and Godron (1984). A lake with minimal shoreline development (i.e., complexity) approximates a circle, but a river or stream shoreline more closely approximates two parallel lines at minimum complexity. The SDR varies with spatial scale. The SDR is calculated by dividing the actual shoreline length L_s within a rectangular moving window by the total length of two sides of the window L_t (i.e., two parallel lines on the right and left banks, which is the theoretical minimum shoreline length; fig. 5–11). The total length of two sides of the window is inputted by the user. The SDR approaches its maximum value (1.0) when the right and left shorelines are linear and parallel as might be expected in a highly channelized system.

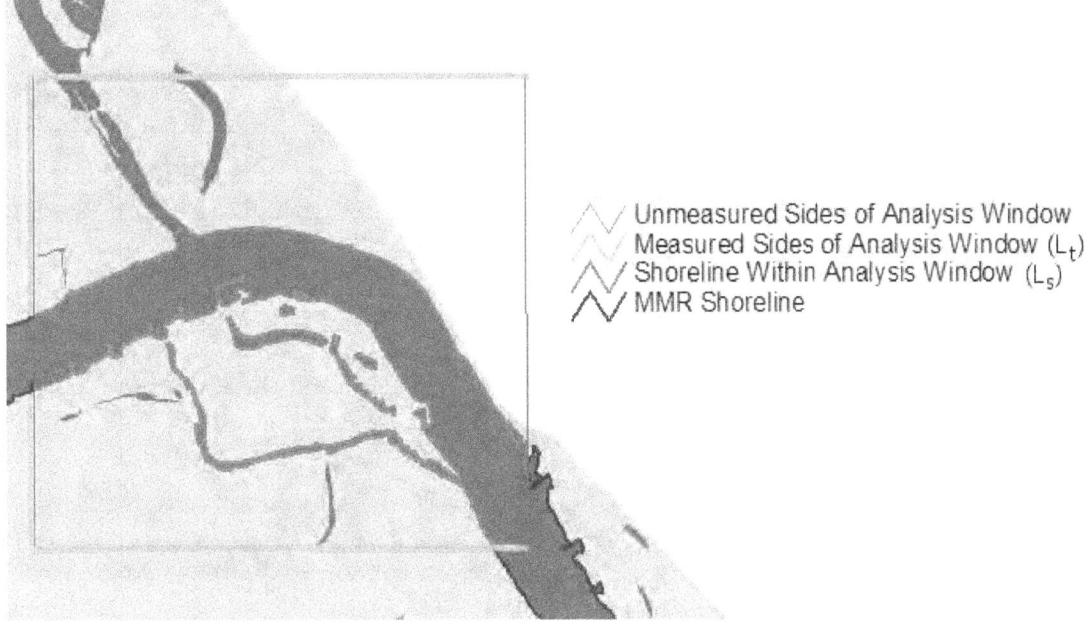

Unmistaken... the legend:
↗↘ Unmeasured Sides of Analysis Window
↗↘ Measured Sides of Analysis Window (L$_t$)
↗↘ Shoreline Within Analysis Window (L$_s$)
↗↘ MMR Shoreline

Figure 5–11. Calculating Shoreline Development Ratio.

1. To start the tool, select **Calculate <u>S</u>horeline Development Ratio by River Mile** from the <u>**Mid_miss_tools**</u> menu options (fig. 5–12).

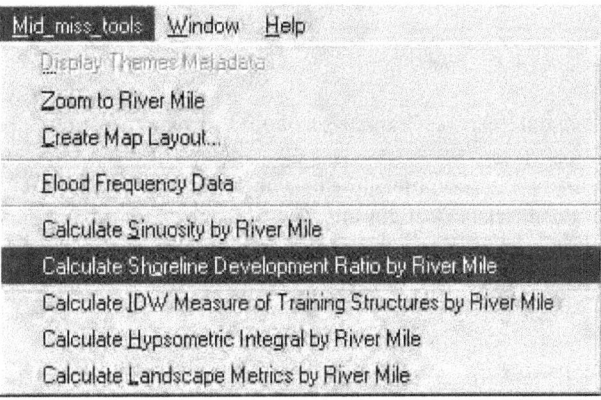

Figure 5–12. Starting Shoreline Development Ratio Tool.

2. Next, **Select Shoreline Development Ratio scale (length of two sides of analysis window):** (fig. 5–13). The first three options have already been preprocessed for the user using the Shoreline Development Ratio Tool. This was done so the user could bypass lengthy processing times and rapidly access these results. The distance between focal points for these scale options was 100 meters.

    a. The first option is *5000 meters (min)*, which was determined to be the minimum value for the Middle Mississippi River because lower values would exclude some shoreline linear features from the analysis.

    b. The second option is *7707 meters **. This value represents 7.5 times the mean active channel width of 1,027.6 meters.

    c. The third option is *10000 meters (max)*. This is the maximum value to use to calculate SDR. Larger values could potentially bias sampling by including shoreline linear features from adjacent meandering sections of the river.

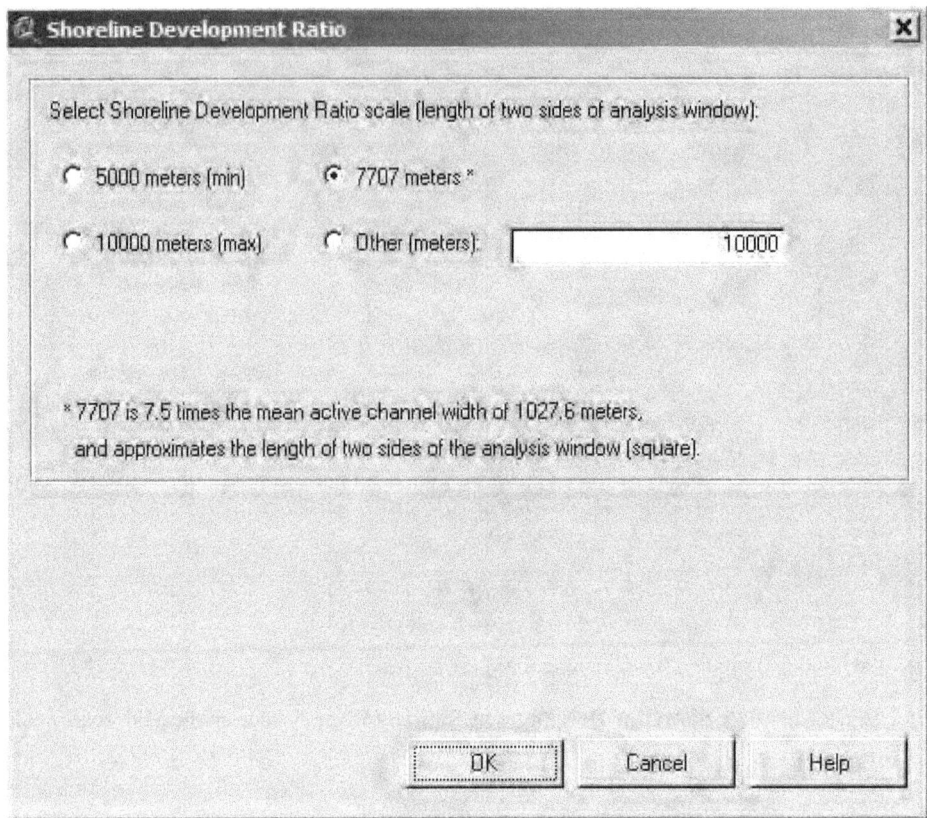

Figure 5–13.   Shoreline Development Ratio input dialog.

    d.   The last option is ***Other (meters)***:. This option allows the user to input their own scale into the equation and also to choose from three different focal point distances (100, 200, and 400 meters). **Selecting this option requires substantially more processing time by the computer**. The user must select an integer value between 5000 (min) and 10000 (max) for scale. The focal point distance options are visible when this scale option is selected.

3.   Click on the **Help** button to display a text file describing the use of the SDR Tool (fig. 5–14).

4.   For this example, select the **preferred scale** option (e.g., ***7707 meters***) and click on the **OK** button (fig. 5–13).

5.   A window opens prompting where on the computer's hard drive to save the outputs. Select a **directory** and click on the **OK** button (fig 5–15).

The output data are saved to the specified directory and three output themes are added to the view (fig. 5–16).

Output themes created by the tool are referenced with the chosen scale (e.g., *Shoreline Development Ratio Points (7707m)*). The first theme depicts the focal points spaced 100 meters apart along the sail line and is attributed with the SDR values calculated at the scale selected (*Sdr*), the total shoreline length within the analysis window for each focal point (*Shorelength*), and the total distance traveled upstream to each individual focal point (*Walkdist*). Second, the tool creates a polygon theme (e.g., *Shoreline Development Ratio by RM (7707m)*) representing river mile segments with each separate polygon attributed by the average SDR value (*Sdr_ave*) of all the focal points that fall within that particular polygon and the river mile label (*River_mile*). Finally, the tool creates a theme, *Shoreline Linears*, representing the Shoreline of the Middle Mississippi River that was used in the analysis. Metadata are created for each theme and are accessible by clicking on the **Display Metadata** button ![M].

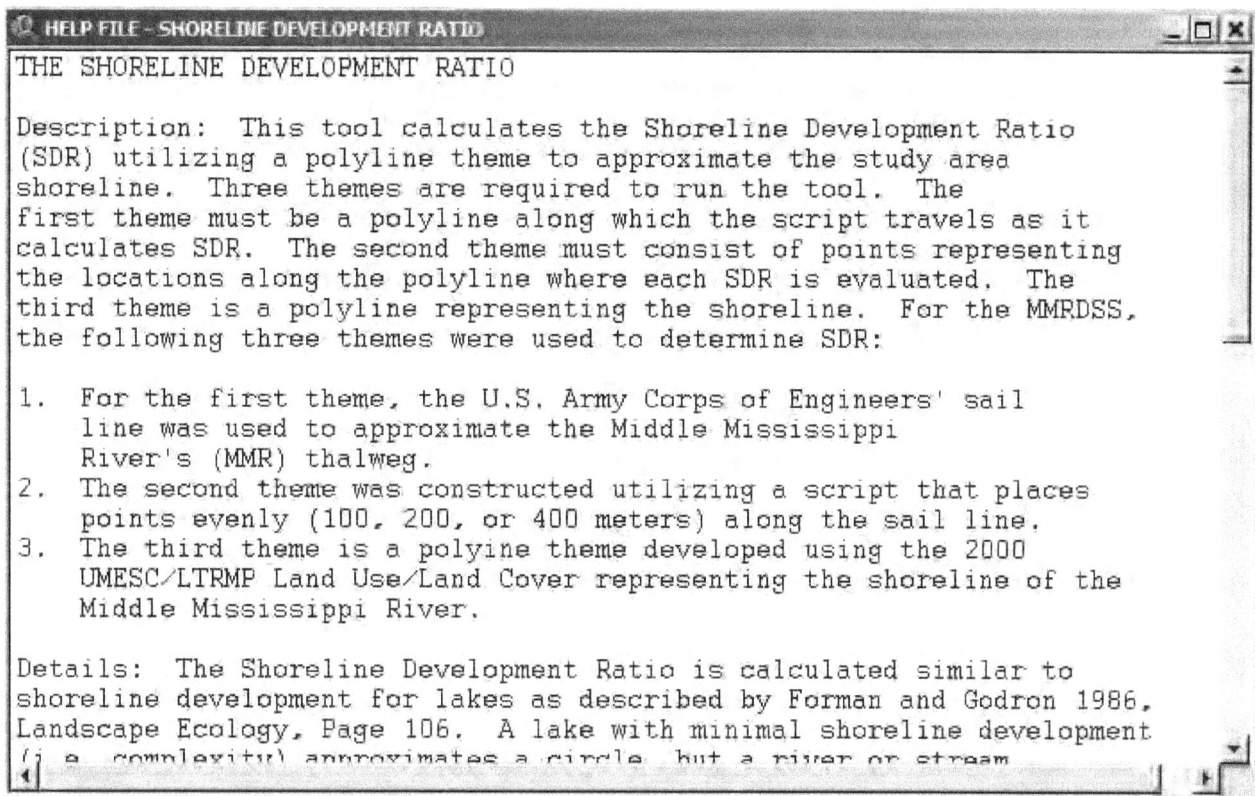

THE SHORELINE DEVELOPMENT RATIO

Description: This tool calculates the Shoreline Development Ratio
(SDR) utilizing a polyline theme to approximate the study area
shoreline. Three themes are required to run the tool. The
first theme must be a polyline along which the script travels as it
calculates SDR. The second theme must consist of points representing
the locations along the polyline where each SDR is evaluated. The
third theme is a polyline representing the shoreline. For the MMRDSS,
the following three themes were used to determine SDR:

1. For the first theme, the U.S. Army Corps of Engineers' sail
    line was used to approximate the Middle Mississippi
    River's (MMR) thalweg.
2. The second theme was constructed utilizing a script that places
    points evenly (100, 200, or 400 meters) along the sail line.
3. The third theme is a polyine theme developed using the 2000
    UMESC/LTRMP Land Use/Land Cover representing the shoreline of the
    Middle Mississippi River.

Details: The Shoreline Development Ratio is calculated similar to
shoreline development for lakes as described by Forman and Godron 1986,
Landscape Ecology, Page 106. A lake with minimal shoreline development
(i.e. complexity) approximates a circle, but a river or stream

Figure 5–14. Shoreline Development Ratio Help File.

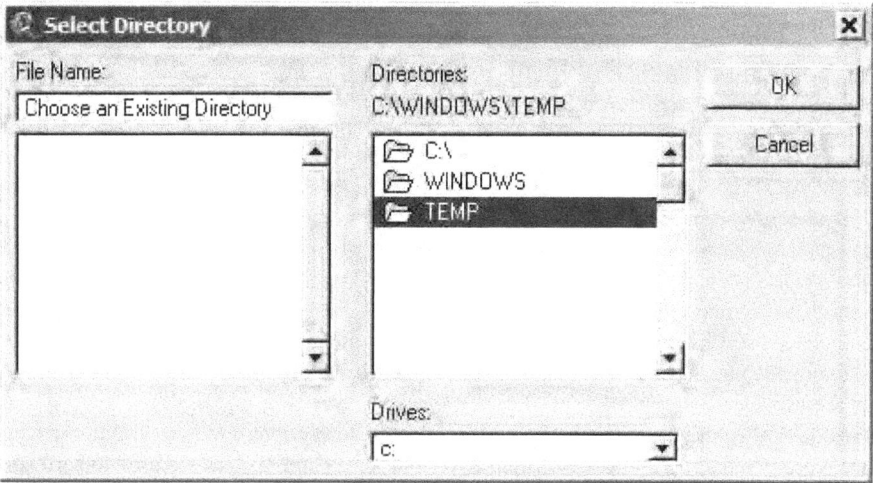

Figure 5–15. Specify Output Directory.

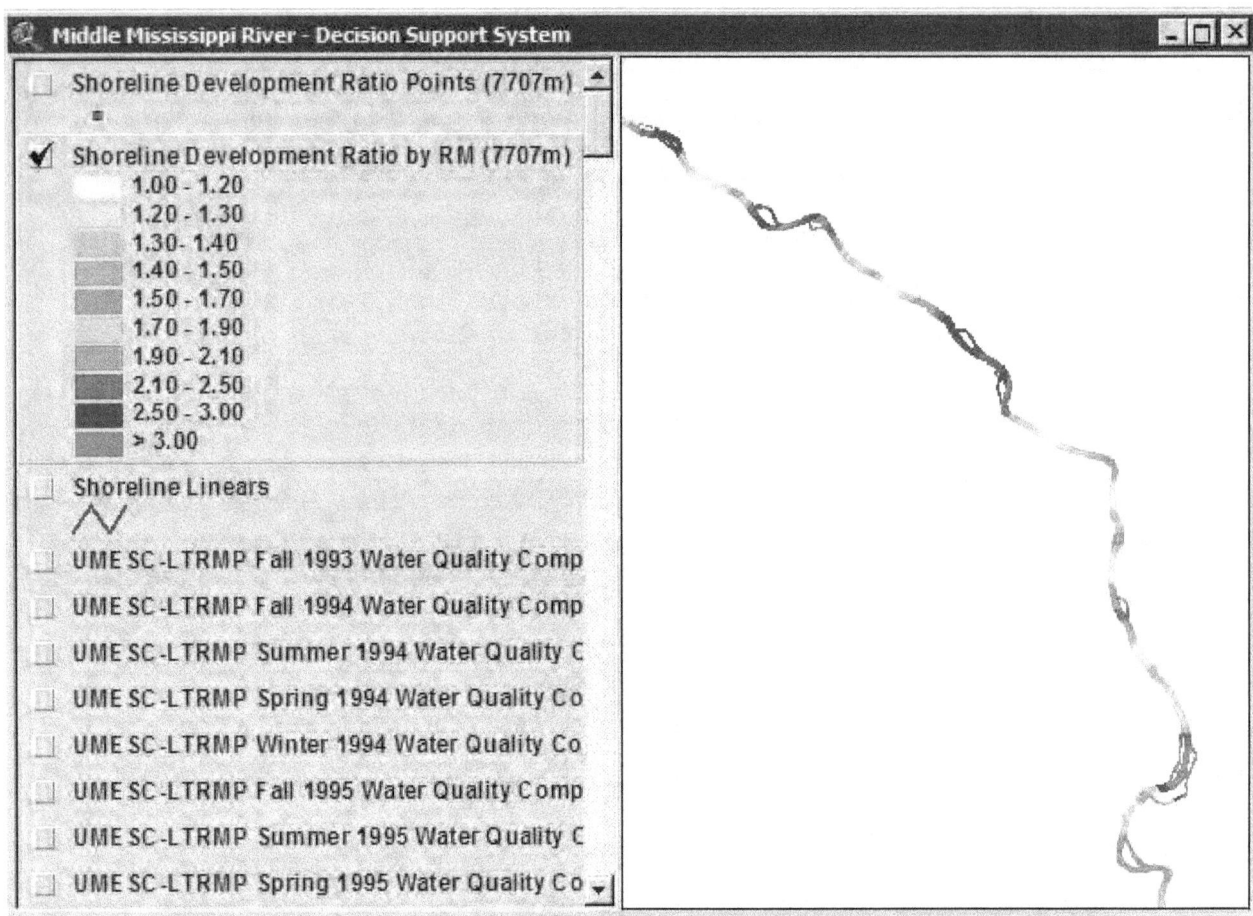

Figure 5–16. Shoreline Development Ratio Outputs.

Inverse Distance Weighted Measure of River Training Structures

Dikes, weirs, and revetments are engineered features designed to maintain the river navigation channel through modification of natural processes such as channel degradation, sediment transport, aggradation, and flooding. However, river training structures may provide physical conditions needed by some aquatic biota. Consequently, biotic abundance and community structure may correlate with proximity to river training structures. However, estimating the influences of river training structures on biota is complex because any given point may be influenced by a number of structures, depending on scale, that vary in length, geometry, and proximity. Inverse distance weighted (IDW) interpolation is a common method for deriving continuous surfaces such as topography or rainfall distribution (Legendre and Legendre, 1998; Lee and Wong, 2001; Lo and Yeung, 2002). Similarly, IDW values were interpolated for the MMRDSS from known lengths of training structures for use in analyses of aquatic organisms. Simply stated, larger structures near the thalweg weigh more heavily (i.e., produce larger values) on IDW values than do smaller structures further away.

This tool calculates several values using the distance from linear features representing river training structures to focal points along a polyline representing the Middle Mississippi River sail line. Three themes are used to perform this analysis within the tool. These themes do not need to be added to the view.

- The first theme is a modified version of the U.S. Army Corps of Engineers' sail line.

- The second theme consists of focal points at regular intervals along the sail line.

- The third theme consists of line features representing dikes, revetments, and weirs acquired from the U.S. Army Corps of Engineers. Revetment line features were derived from the existing revetment polygon theme using a tool to approximate polygon centerlines. Revetment, dike, and weir polylines were finally merged into a single theme.

1. To start the tool, select **Calculate IDW Measure of Training Structures by River Mile** from the **Mid_miss_tools** menu options (fig. 5–17).

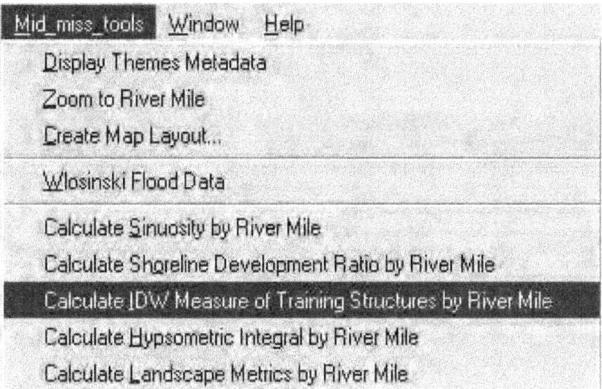

Figure 5–17.   Starting Inverse Distance Weighted Measure of Training Structures by River Mile Tool.

2. Next, **Select IDW measure of training structures scale (radius of analysis circle):** (fig. 5–18).
    a. The first option is *1000 meters (min)*, which was determined to be the minimum value. Lower values would exclude some channel training structures in the Middle Mississippi River from the analysis.
    b. The second option is *5332 meters ***. This value was derived from a circle having an area equal to the mean area of adjacent MMR 11 and 14 digit hydrologic unit code (HUC) watershed polygons.
    c. The third option is *10000 meters (max)*. This is the maximum value to use to calculate IDW. Larger values could potentially bias sampling by including some training structures from adjacent meandering sections of the river.

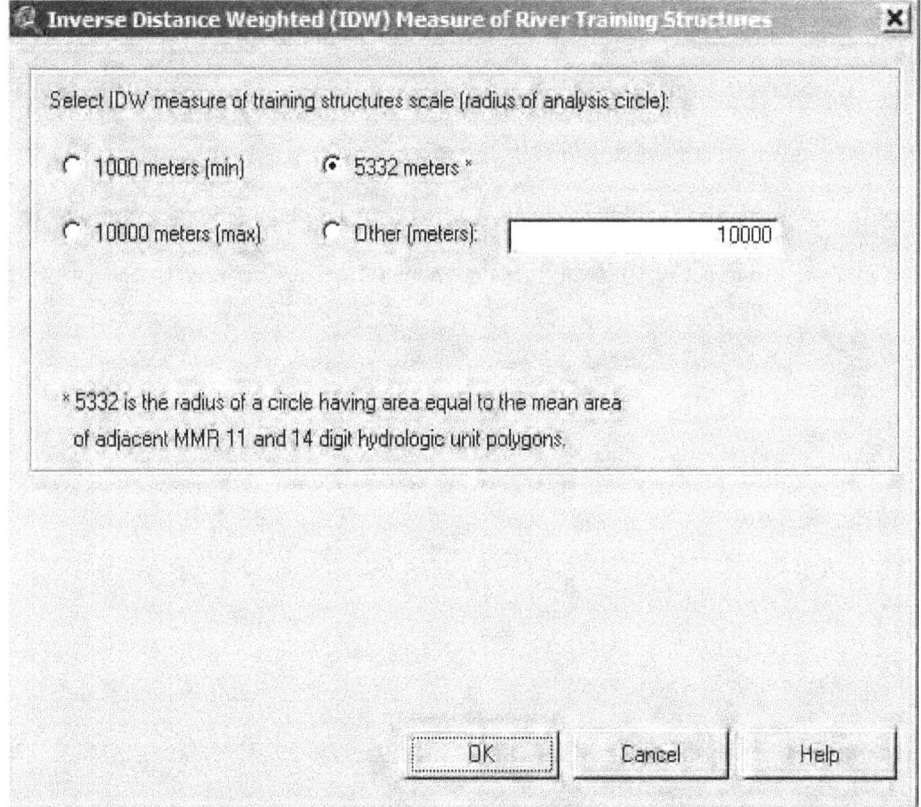

Figure 5–18.   Select Appropriate Scale Value.

    d. The last option is **Other (meters):**. This option allows the user to input their own scale into the equation and also to choose from three different focal point distances (100, 200, and 400 meters). **Selecting this option requires substantially more processing time by the computer**. The user must select an integer value between 1000 (min) and 10000 (max) for scale. The focal point distance options are visible when this scale option is selected.

3. Click on the **Help** button to display a text file describing the use of the IDW Tool (fig. 5–19).
4. Select the **preferred scale** option (e.g., *5332 meters **) and click on the **OK** button (fig. 5–18).

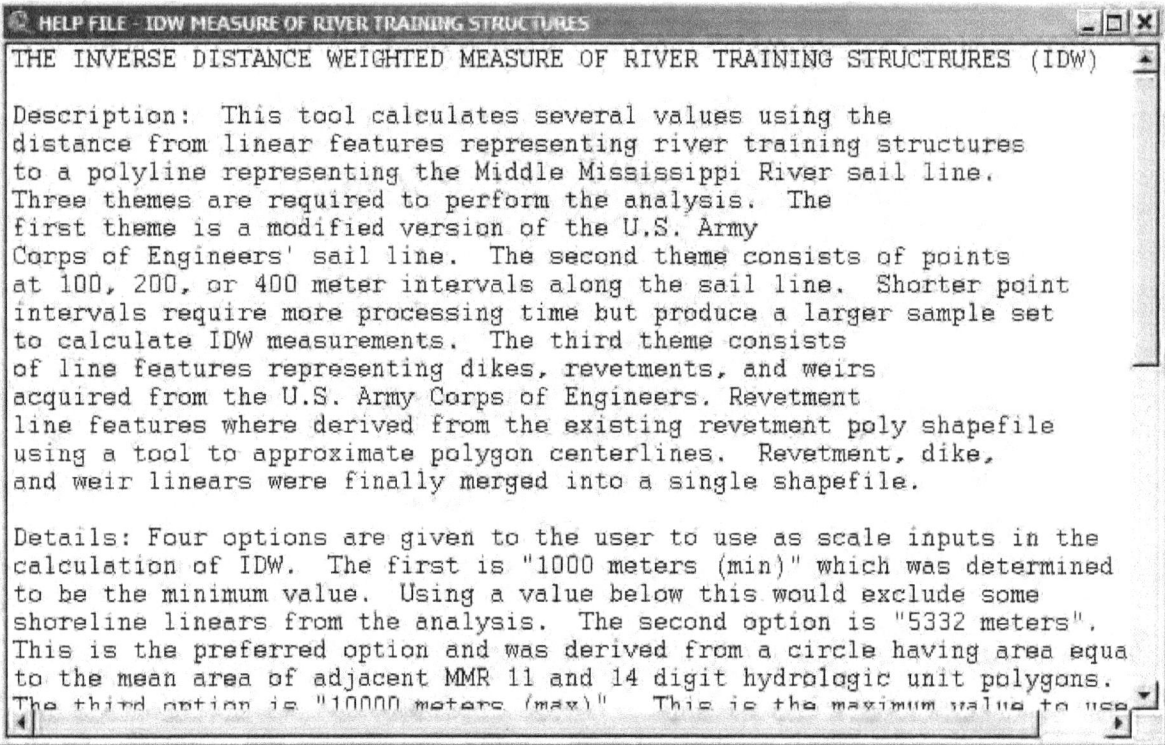

    Figure 5–19. Inverse Distance Weighted Measure of Training Structures Help File.

5. A window opens prompting where on the computer's hard drive to save the outputs. Select a **directory** and click on the **OK** button (fig. 5–20).
    The output data are saved to the specified directory, and three output shapefiles are added to the view (fig. 5–21).

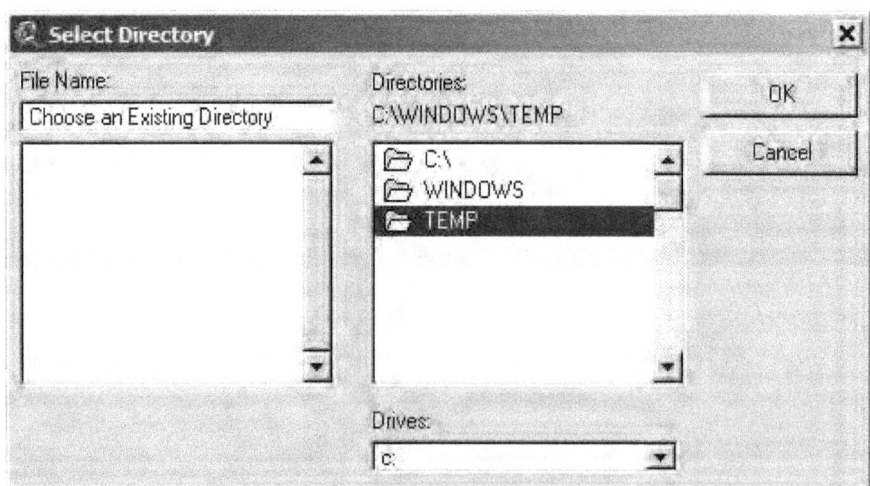

    Figure 5–20. Specify Output Directory.

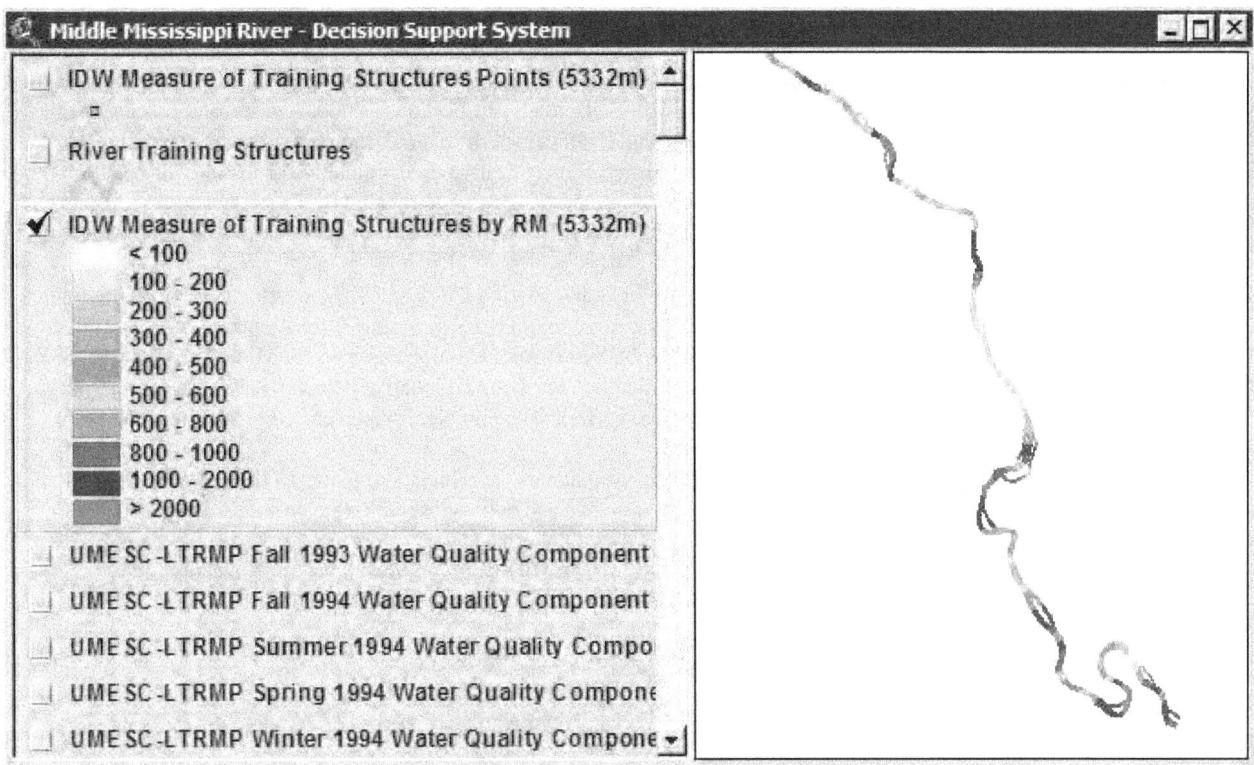

Figure 5–21. Inverse Distance Weighted Measure of River Training Structures Outputs.

This tool creates three themes. The first theme (e.g., *IDW Measure of Training Structures Points (5332m)*) represents focal points spaced 100 meters apart along the sail line and its associated attribute table. The first attribute in the table (*Sumlinelng*) is the total length of linear features within the circular buffer around each focal point.

The second attribute (*Lngth2mind*) is the sum of each feature's length divided by the squared minimum distance between the feature and the focal point of the moving window.

The third attribute (*IDW*) is the Inverse Distance Weight of the features in each sampling window (IDW = Sum[w*d^2]/ Sum[1/d^2]) where weight or w = linear length of each feature and d = minimum distance between each feature and the center of the search window.

The fourth attribute is the total distance traveled upstream to each individual focal point (*Walkdist*). The second theme created (e.g., IDW *Measure of Training Structures by RM [5332m]*) represents river mile segments as separate polygons that are attributed by the average IDW value (Idw_ave) for all of the focal points falling within that particular river mile, and the river mile number for each polygon (*River_mile*).

Finally, the tool creates a theme, *River Training Structures*, representing the River Training Structures of the Middle Mississippi River used in the analysis. Metadata are created for each theme and are accessible by clicking on the **Display Metadata** button $\boxed{\text{M}}$.

Hypsometric Integral

Hypsometry provides a means to describe the variability in topography of watersheds and water bodies derived from a frequency distribution of elevations (or depth). For river reaches, hypsometric curves for bathymetry can be generated as plots of normalized depth against normalized cumulative area. Normalizing (adjusting the values to a common value) the depth and area allows the user to compare results between different river miles or river reaches on the basis of a common standard. The hypsometric integral (HI) represents total area under a hypsometric curve (Harlin, 1978; Luo, 1998). Total area within the reach serves to normalize area at or below a given depth. Conversely, maximum reach relief normalizes depth on the y-axis. When applied as a general descriptor of reach bathymetry, low values of HI are expected in highly channelized reaches whereas higher values are expected in aggrading reaches (Hurtrez and others, 1999).

Although the percent hypsometric curve and HI have been successfully used to characterize fluvial geomorphology within watersheds (Hurtrez and Lucazeau, 1999; Hurtrez and others, 1999; Haltuch and Berkman, 2000; Awasthi and others, 2002), comparisons among basins or water bodies with unequal area and relief can be difficult. A more robust characterization of small basins was offered by Harlin (1978), who suggested treating the hypsometric curve as a cumulative probability distribution by fitting a third order polynomial to the percent hypsometric curve and integrating to obtain HI. Then, skewness and kurtosis can be used to quantitatively describe the curve. Together, these traits better characterize the curve and reduce the confounding effects of relying solely on HI.

Luo (1998) modified Harlin's (1978) automated approach of characterizing the percent hypsometric curve for GIS applications using ARC/INFO's ARC macro language (AML) and executable FORTRAN programs (ftp://ftp.iamg.org/VOL24/v24-8-10.zip). These programs were modified to accommodate MMR bathymetric data and to characterize MMR hypsometric curves for the MMRDSS. Program results of polynomial fit were verified by regressing cumulative percent area and percent depth coordinate pairs with the general linear model (GLM) procedure in SAS. The HI values were verified by coding trapezoidal approximation of the integral within the FORTRAN program.

Bathymetry raster grids representing depth (only grid cells <= 0 meters were included) at the 10 percent exceedance levels were used in the hypsometric analysis. For the MMRDSS, polynomials and curve attributes were processed for River Mile (RMs) 28 through 80 and the entire Open River Trend Analysis Area (TAA).

1. To start the tool, select **Calculate Hypsometric Integral by River Mile** from the **Mid_miss_tools** menu options (fig. 5–22).

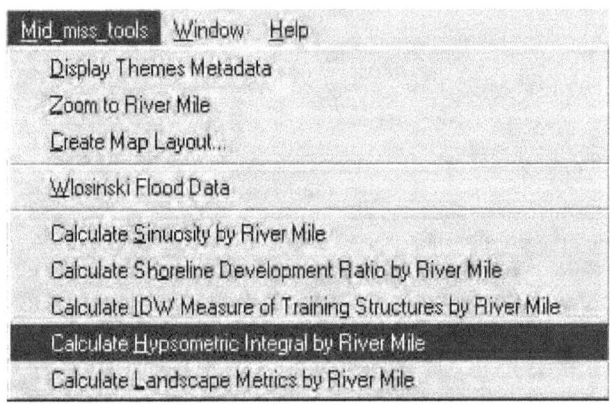

Figure 5–22. Starting Hypsometric Integral Tool.

2. Next, select how the HI Tool will normalize output (fig. 5–23). Two options are available.
   a. The first option, *Calculate Hypsometric Integral (Normalize by River Mile),* uses the total area and maximum depth within each river mile segment as the normalizing values on the x and y axes of the percent hypsometric curve from which the HI is obtained. These values are different for each river mile.
   b. The second option, *Calculate Hypsometric Integral (Normalize by Trend Analysis Area)*, indicates that the percent hypsometric curve and subsequent HI for each river mile were developed using total area and relief within the TAA bathymetry grid as normalizing (denominator) values on the curve's x and y axes. This option may better standardize HI values for comparing diverse river reaches.
3. Click on the **Help** button to display a text file describing the use of the HI Tool (fig. 5–24).
4. Select **Calculate Hypsometric Integral (Normalize by River Mile)** and click on the **OK** button (fig. 5–23).
5. A window opens prompting where on the computer's hard drive to save the outputs. Select a **directory** and click on the **OK** button (fig 5–25).

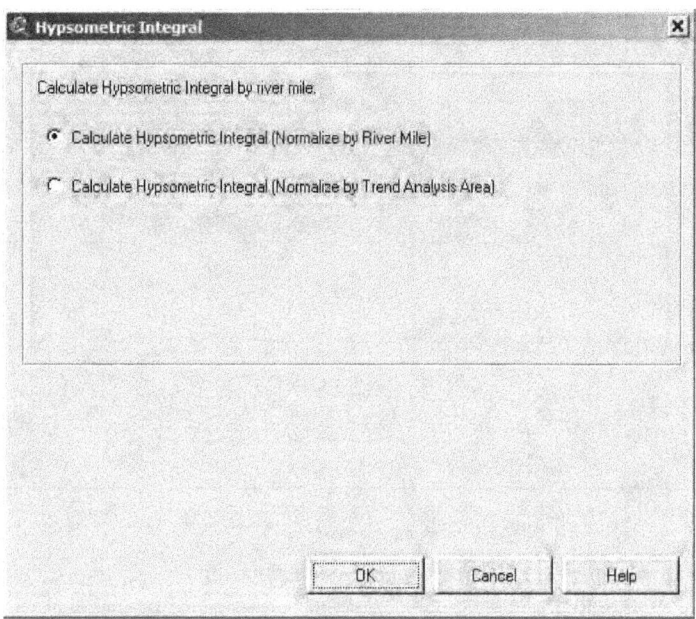

Figure 5–23.   Select Hypsometric Integral Output.

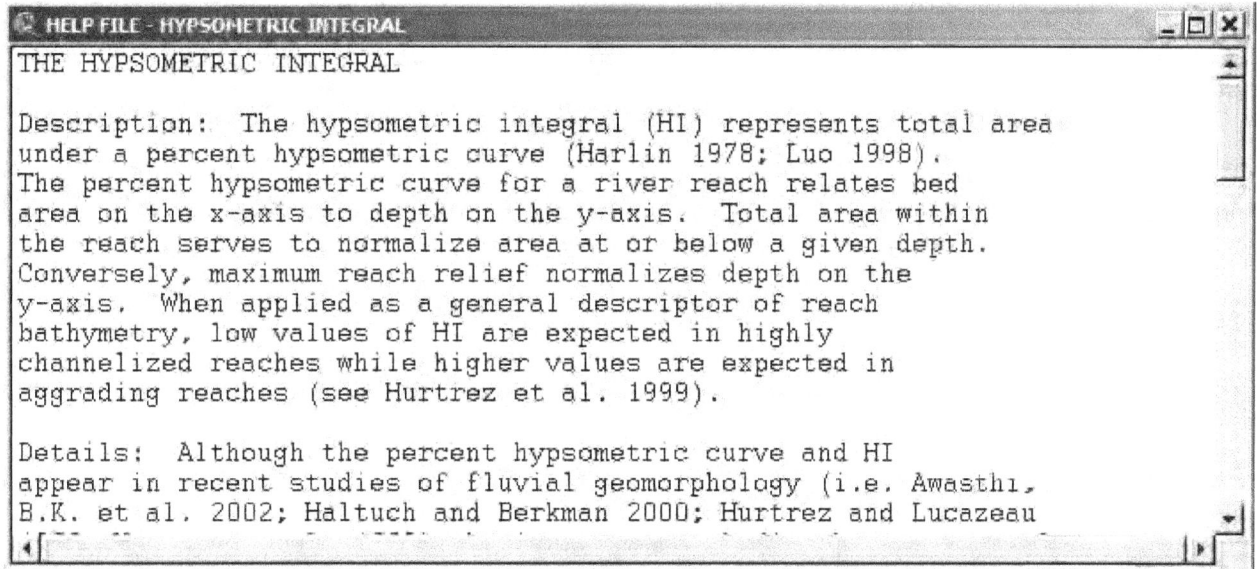

Figure 5–24.   Hypsometric Integral Help File.

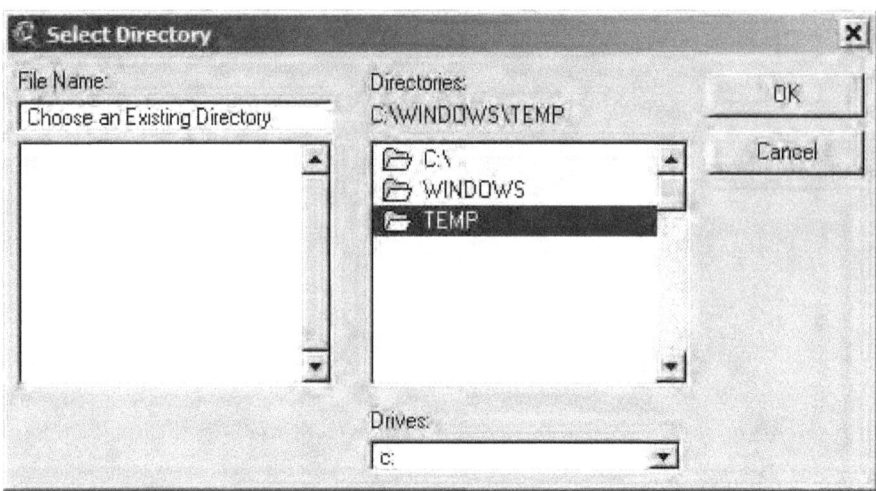

Figure 5–25.   Specify Output Directory.

The output data are saved to the specified directory and one output shapefile is added to the view (fig. 5–26).

These products and the tools used to derive them may assist river investigators in a number of ways. For example, the literature proposes that HI discerns aggrading reaches from degrading reaches. Moreover, HI may be useful for evaluating connectivity and physical processes (e.g., water movement) as well as affinities of species or guilds for particular bathymetric profiles. The AML scripts and compiled FORTRAN programs allow investigators to quantify differences in HI among other reaches of the river. The products offer the ability to uniquely characterize and relate the percent hypsometric curve to riverine biota and ecological processes. Metadata are created for each theme and are accessible by clicking on the **Display Metadata** button ⊞.

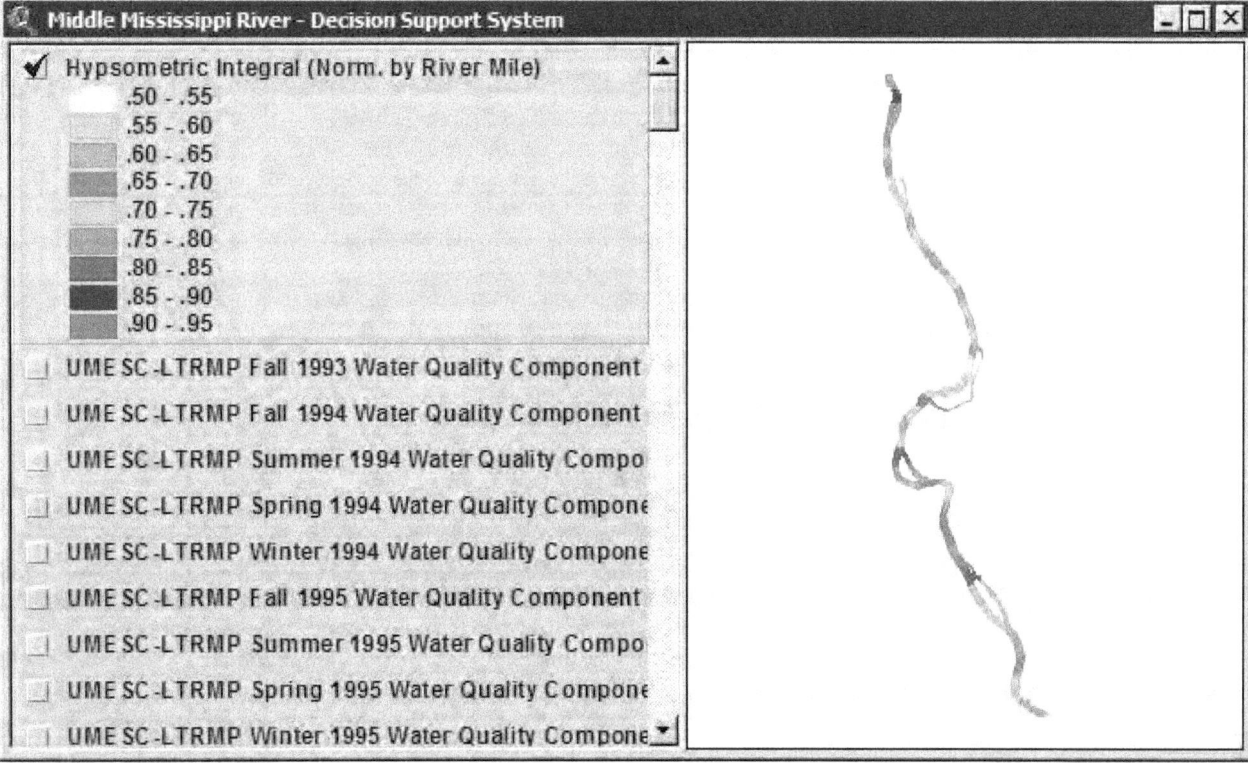

Figure 5–26.   Hypsometric Integral Output.

Fragstats Version 3.3 Landscape Metrics

FRAGSTATS Version 3.3 is a software application developed to quantify habitat metrics for landscape ecology at the patch, class, and landscape levels (McGarigal and others, 2002). For the MMRDSS, these metrics were derived from the U.S. Army Corps of Engineers (COE) aquatic area raster grids and bathymetry grids of depth <= 0 meters at the 10 percent discharge exceedance level. For each grid, 101 landscape metrics (Appendix 3) were calculated for the entire TAA (RM 28–80). The MMRDSS users should visit the FRAGSTATS Web site at http://www.umass.edu/landeco/research/fragstats/fragstats.html to obtain information about the software and detailed descriptions of individual metrics.

Once processed, dBASE files are created and imported as tables into ArcView. Landscape level metrics processed by river mile are joined to a shapefile representing river mile polygons.

1. To start the tool, select **Calculate Landscape Metrics by River Mile** from the **Mid_miss_tools** menu options (fig. 5–27).

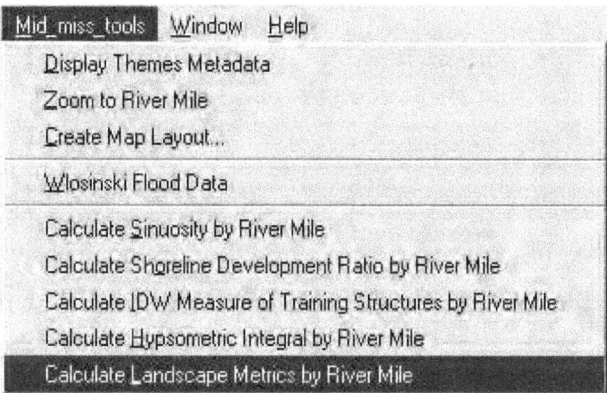

Figure 5–27.   Starting Landscape Metrics Tool.

2. Next, select the input theme to Calculate Landscape Metrics (fig. 5–28). Options include *Calculate Landscape Metrics Using UMRS Bathymetry* or *Calculate Landscape Metrics Using COE Aquatic Areas*.

3. Click on the **Help** button to display a text file describing the use of the Landscape Metrics Tool (fig. 5–29).

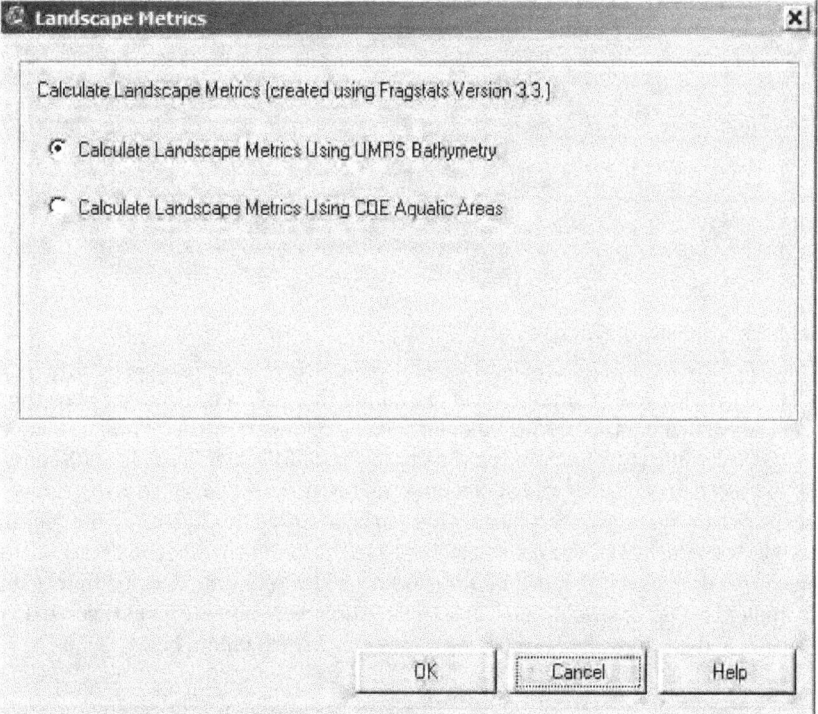

Figure 5–28.   Select Landscape Metrics Output.

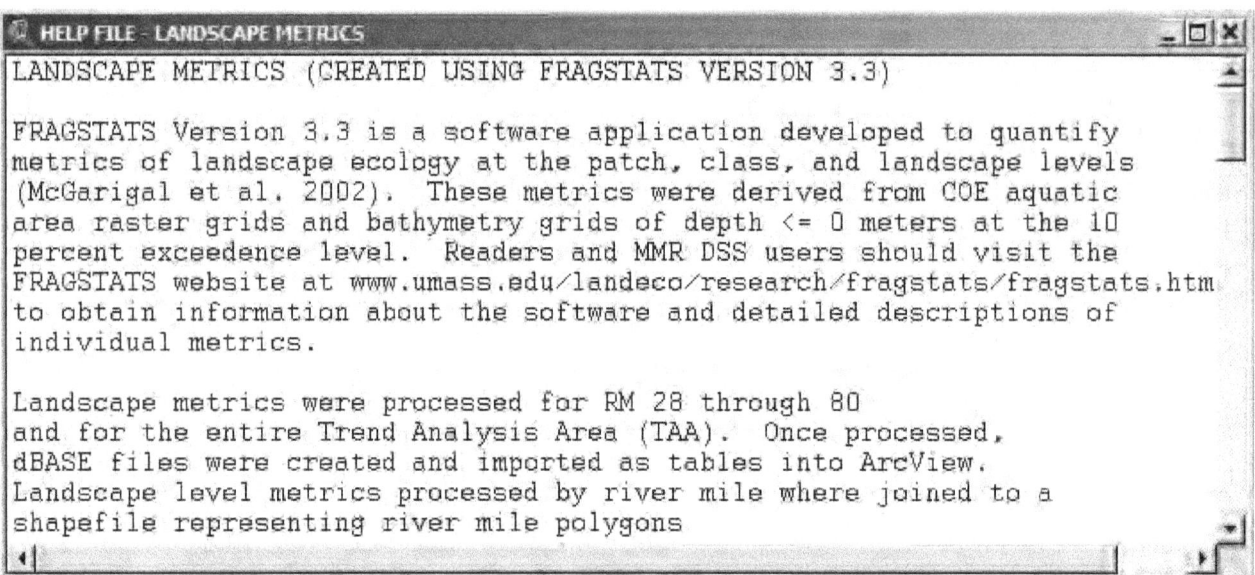

Figure 5–29. Landscape Metrics Help File.

4. Select the option ***Calculate Landscape Metrics Using UMRS Bathymetry*** and click on the **OK** button (fig. 5–28).
5. A window opens prompting where on the computer's hard drive to save the outputs. Select a **directory** and click on the **OK** button (fig. 5–30).

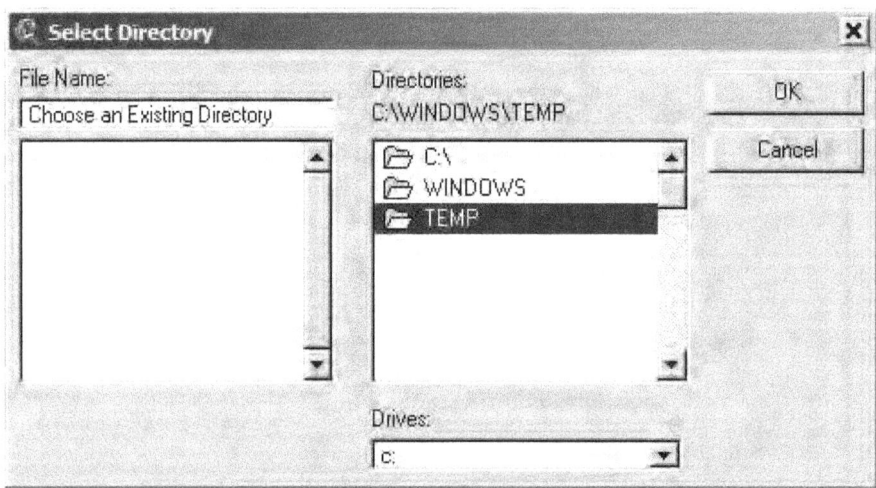

Figure 5–30. Specify Output Directory.

The output data are saved to the specified directory, and one output shapefile is added to the view (fig. 5–31).

The shapefile subsequently added is attributed with the full suite of landscape metrics within the associated attribute table, and the theme itself is shaded according to its Simpson's Diversity Index (Sidi) value. The value of Simpson's Diversity Index represents the probability that any two pixels selected at random would be different patch types (e.g., polygons of depth strata). The entire suite of landscape metrics for each river mile polygon can be accessed by clicking on the **Identify Tool** and selecting a feature within the view (fig. 5–31). The theme can be shaded according to other landscape metrics by selecting **Edit Legend...** from the **Theme** menu options. The Legend Editor appears and the user can select a different attribute to shade the theme using the **Classification Field** drop-down menu. Click on the **Apply** button to accept changes and view the theme. Metadata are created for each theme and are accessible by clicking on the **Display Metadata** button.

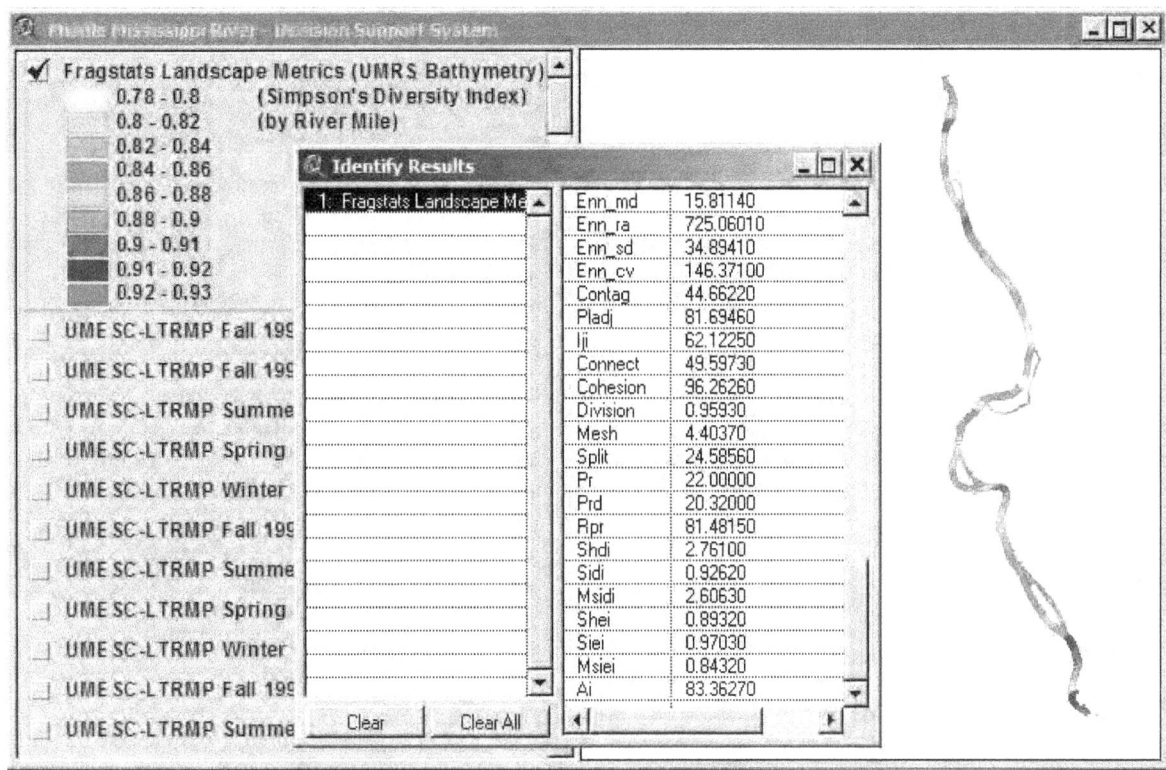

Figure 5–31. Landscape Metrics Output.

Section 6: Clipping Tool

Background

The Clipping Tool was first developed as an ArcView extension provided with the Habitat Needs Assessment (HNA) GIS Query Tool for the Upper Mississippi River System (DeHaan and others, 2000). The Clipping Tool allows the user to clip data themes and view summary statistics of the clipped region. What makes the Clipping Tool a powerful analytical tool is its ability to clip through multiple themes and compare those themes on a common attribute.

The Clipping Tool consists of a **Clip With Area Tool** and the **Clip With Theme** button. Before either can be used, the View's map units must be set (if not already set). The **View** Units are set by selecting **Properties** from the **View** menu options. Once the dialog window opens, set the **Map Units: to meters** for the MMRDSS.

The MMRDSS uses several land cover/land use themes as input with the Clipping Tool. These themes consist of the *2000 LTRMP Land Cover/Land Use, 1989 LTRMP Land Cover/Land Use, 1975 LTRMP Land Cover/Land Use,* and the *1890's Mississippi River Commission LCLU.* **It is important to note that these individual data themes were not all originally interpreted according to the same standard operating procedures and also were not all interpreted at the same scale.** These themes' attribute tables and corresponding fields were cross-walked by the Upper Midwest Environmental Sciences Center (UMESC) so that comparisons could be made using fields shared by all of the themes. The *2000 LTRMP Land Cover/Land Use* theme was used as the model. These themes were added to a separate view within the MMRDSS called *Middle Mississippi River - Land Cover Clipping Themes.* Use this view as a starting point for analysis using the Clipping Tool.

Clipping Multiple Themes Using a User-Defined Polygon

A user-defined polygon can be used to clip one or multiple data themes. In this example, the user zooms into the area surrounding Marquette Island, near RM 49 and clips the area surrounding the island.

1. Make the themes *2000 LTRMP Land Cover/Land Use (2)*, *1989 LTRMP Land Cover/Land Use (2)*, *1975 LTRMP Land Cover/Land Use (2)*, and *1890's Mississippi River Commission LCLU (2)* active (make multiple themes active by selecting one theme in the table of contents and selecting the rest with the **Shift** key depressed).
2. Once the themes are selected, press the **Clip With Area Tool** .
3. Next, define the area of analysis by **clicking** within the view **once** with the **left mouse button** to start the polygon and **press again** for each vertex in the polygon. **Double-click the mouse** to finish the polygon (fig. 6–1) and display the Clipping Tool input window.

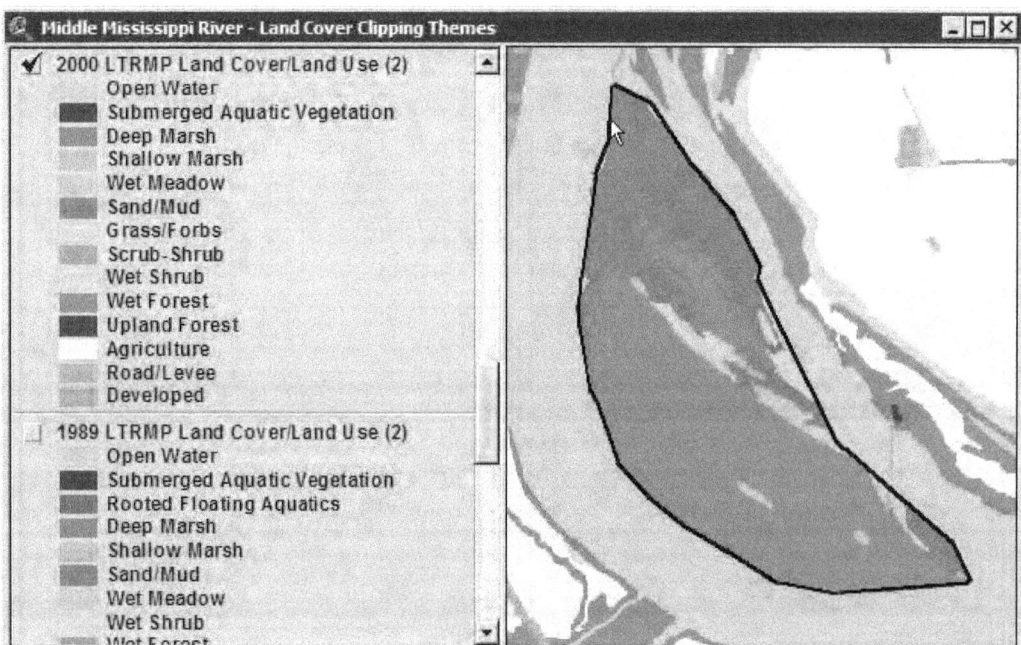

Figure 6–1.   Use the Clipping Tool to create a clipping polygon.

4. Next, Fill in Output Specifications within the **Summarize by User Polygon** dialog window (fig. 6–2). This dialog window has user input areas to do the following:
    a. Specify a **Basename for Output Docs** created by the tool (for this example, type in **Marquette Island**).
    b. Select an **Output Directory** to place generated files.
    c. **Select Field to Summarize** common to all active themes on which summary statistics will be performed from the drop-down menu (for this example, select **Class_15**).
    d. Select the type of **Classification of Table/Theme** used with the tables and themes from the drop-down menu (for this example, select **Unique Values**). If the field being summarized is a character string, the user will only be able to designate the *Classification of Table/Theme* as **Unique Values** and the *Number of Classes* option will be deactivated.
    e. Check whether to **Create Chart(s)** as an output or whether to **Create Theme from the Clip Polygon**. If the field being summarized is numerical, the dialog box also prompts the user to specify the **Number of Classes** to use in grouping the resulting numbers.
    f. After filling in the information, click the **OK** button in the dialog box, the Clipping Tool will use the clipping polygon to extract data from the active themes and create new graphic and tabular outputs.

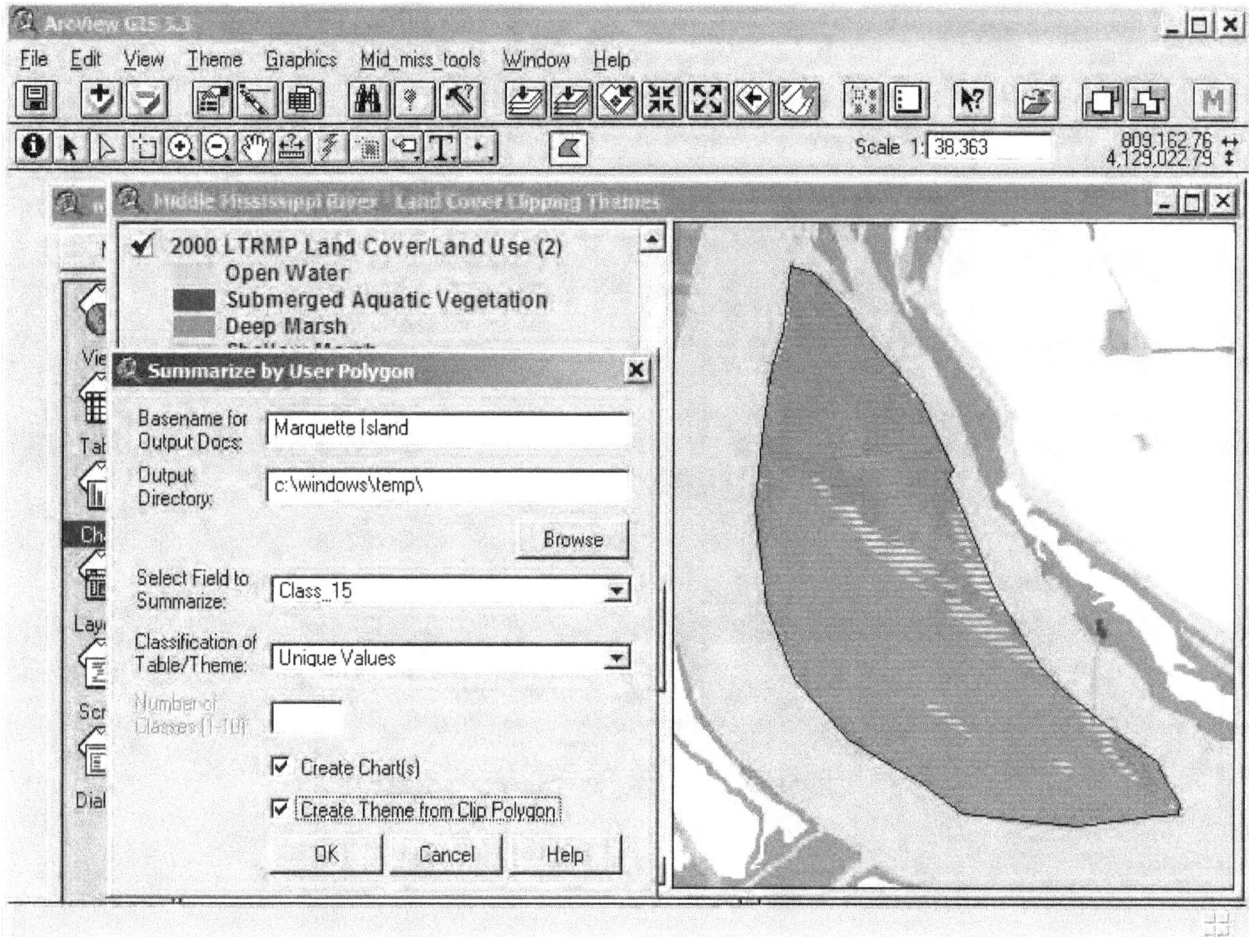

Figure 6–2.   Dialog box used to designate output specifications for the Clipping Tool.

5. Once the process is finished, the areas from the active themes that overlapped the clipping polygon are added to the active View window (fig. 6–3). If an input theme does not overlap the designated clipping polygon, no output theme will be created for that particular theme. The associated tables and charts are also added to the project and can be accessed from ArcView's Project window (fig. 6–4).

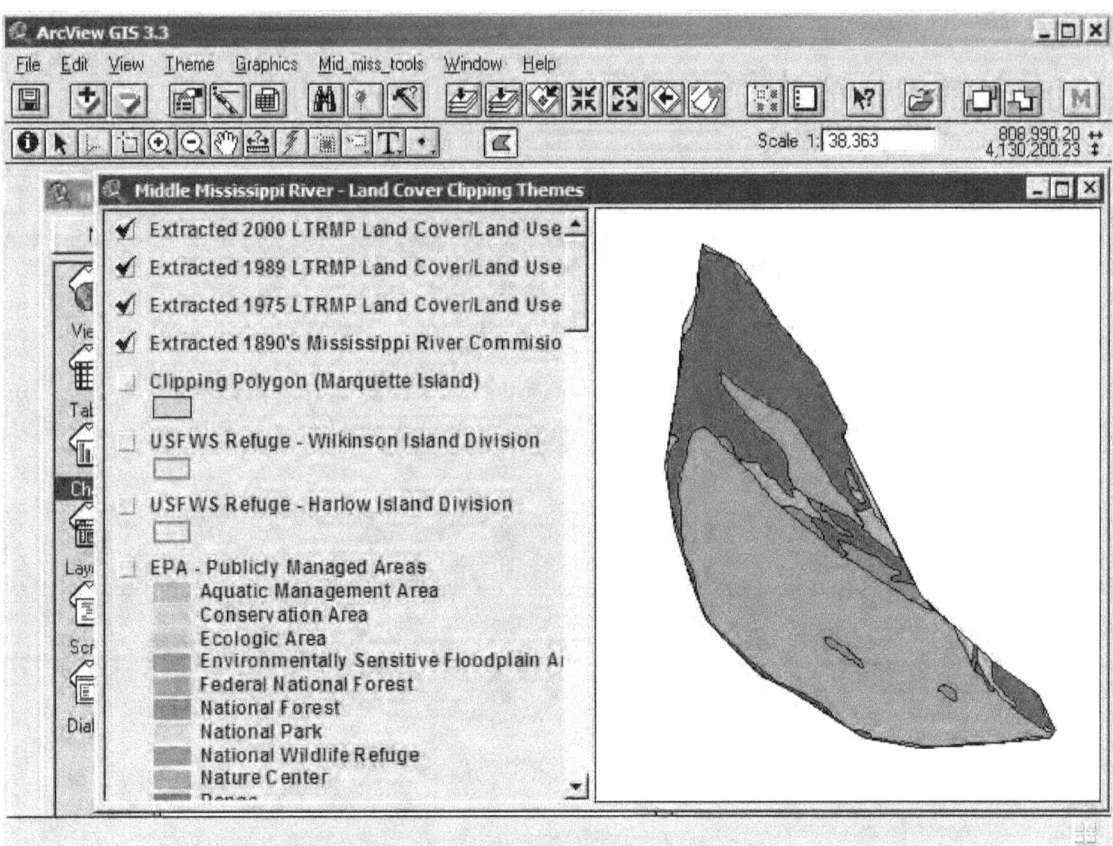

Figure 6–3. Extracted files drawn in the View window after the clipping procedure.

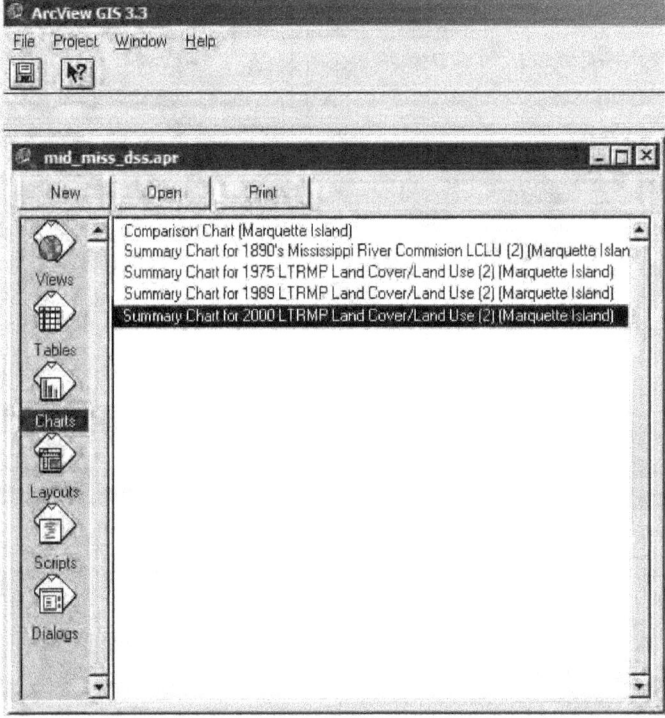

Figure 6–4. Charts created when the example clipping procedure is completed.

6. To open up a specific chart, **select it** within the Project window and click on the **Open** button. Select the chart labeled *Summary Chart for 2000 LTRMP Land Cover/Land Use (2) (Marquette Island)* (fig. 6–4). All the charts created have the area calculation along the y-axis (fig. 6–5). If the map units were specified in the View properties dialog box before running the Clipping Tool, acres are calculated for the output themes. Acres will also be used in the summary statistics. The classes that the data were summarized on are located along the x-axis of the chart. A chart is created for each data theme clipped. A comparison chart is also created so the user can identify trends in the data (fig. 6–6). Note: If the user opens a chart and gets a message that says, "There is not enough space to plot the chart…" just increase the size of the window and the chart will display. If the chart still does not display, there are too many columns in the chart to view. If this happens, results will need to be compared solely within the summary tables.

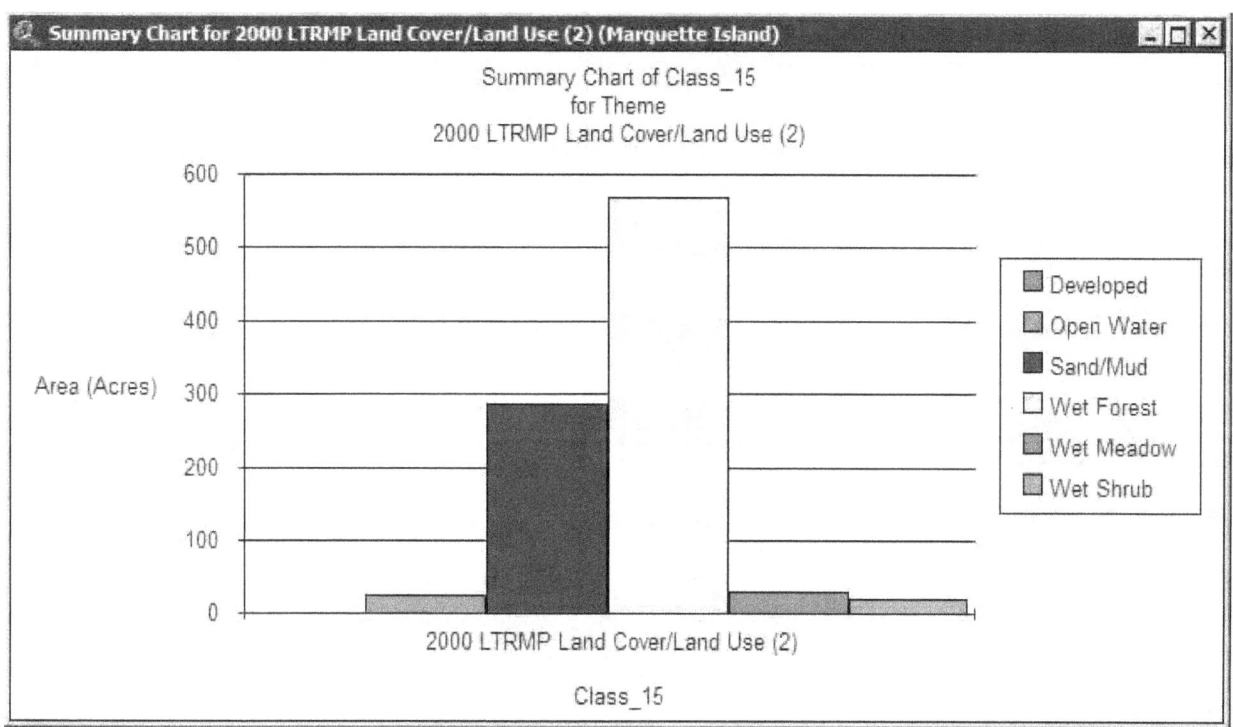

Figure 6–5. Summary chart for one of the clipped themes.

7. The charts can be altered to look at the trends between different classes. To do this, select the **Chart Tool** button (fig. 6–6). Pressing this button opens a dialog box that allows the user to select any combination of available classes and regraph the data showing only those classes. Hold the **Shift** key to **select multiple classes**.

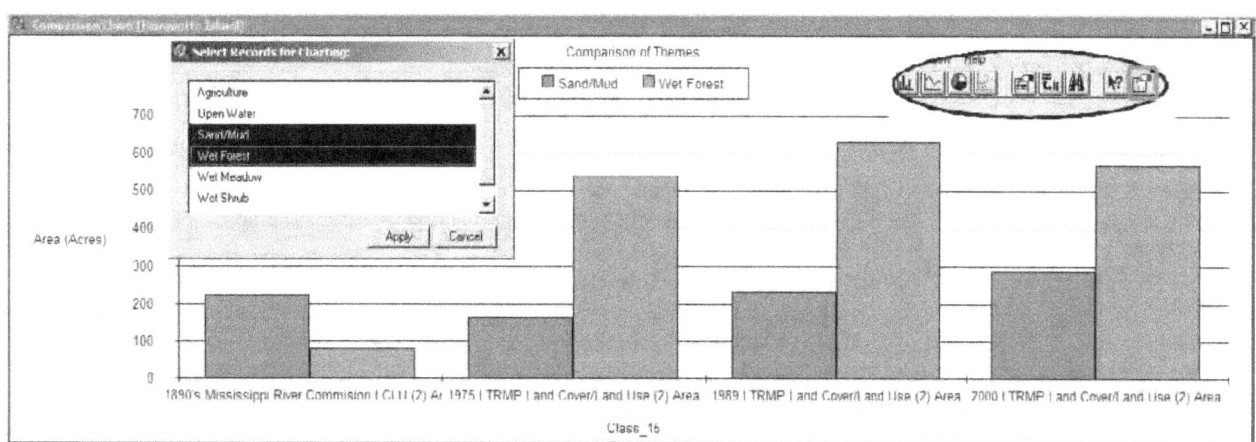

Figure 6–6. Comparison chart for all clipped themes showing the Sand/Mud and Wet Forest classes.

8. Besides creating charts with the Clipping Tool, summary tables are automatically created for each theme clipped (fig. 6–7). Tables for the individual themes have fields for the class description, the number of polygons for each class description, the total acres of that class within the clip area, and the percentage of that class within the clip area.

Class_15	Count	Area (Acres)	Percent
Developed	1	1.14	0.1
Open Water	26	25.61	2.7
Sand/Mud	6	287.54	30.7
Wet Forest	2	568.87	60.7
Wet Meadow	7	31.35	3.3
Wet Shrub	3	22.54	2.4
TOTAL	45	937.06	100.0

Figure 6–7. Summary table for one of the clipped themes.

9. If multiple data themes are clipped, a comparison table is also created (fig. 6–8). This table has a field for the common class description and a field for each data theme that represents the total area of each class type.

Comparison Table (Marquette Island)

Class_15	Mississippi River Commission LCLU	1975 LTRMP Land Cover/Land Use (2) Area	LTRMP Land Cover/Land Use (2) A	2000 LTRMP Land Cover/Land Use (2) Are
Agriculture	0.38	0.00	0.00	0.00
Open Water	632.83	195.88	44.98	25.61
Sand/Mud	223.48	165.83	232.50	287.54
Wet Forest	80.37	541.25	630.81	568.87
Wet Meadow	0.00	11.12	10.83	31.35
Wet Shrub	0.00	22.99	11.58	22.54
Developed	0.00	0.00	6.35	1.14

Figure 6–8. Comparison table for all clipped themes showing acreages for each class.

Clipping Multiple Themes Using an Existing Polygon

Besides making a polygon interactively to clip themes, the Clipping Tool also provides the option to clip themes by using an existing data theme. The user may clip with an entire data theme or a subset of a data theme using selected polygons, for example:

1. Make the theme *USFWS Refuge - Harlow Island Division* active. Zoom to the extent of that theme using the **Zoom to Active Theme** button ![button].
2. Next, make the themes *2000 LTRMP Land Cover/Land Use (1)*, *1989 LTRMP Land Cover/Land Use (1)*, *1975 LTRMP Land Cover/Land Use (1)*, and *1890's Mississippi River Commission LCLU (1)* active (make multiple themes active by selecting one theme in the table of contents, then select the rest with the **Shift** key depressed). Make sure the theme *USFWS Refuge - Harlow Island Division* is **NOT** active since this theme will be used to clip the other themes.
3. Next, press the Clipping Tool **(Clip with Theme)** button ![button]. This brings up the **Load Polygon from Theme** dialog box (fig. 6–9). In this dialog box, select the name of the theme to use as the bounding polygon for clipping the other theme(s) that are active. Once the user has selected the data theme, click on the **Load** button. For this example, select the theme *USFWS Refuge – Harlow Island Division*.
4. The Summarize by User Polygon dialog window opens. This window is the same as the one used when the user clipped with an interactively created polygon (fig. 6–2). Fill in the input options the same as in Figure 6–2, except name the **Basename for Output Docs** Harlow Island NWR. This operation generates products similar to the previous example (i.e., themes, charts, and tables). Figure 6–10 displays an example of output generated by the Clipping Tool using the data theme shown in Figure 6–9. Open the output charts and tables using the Project window.

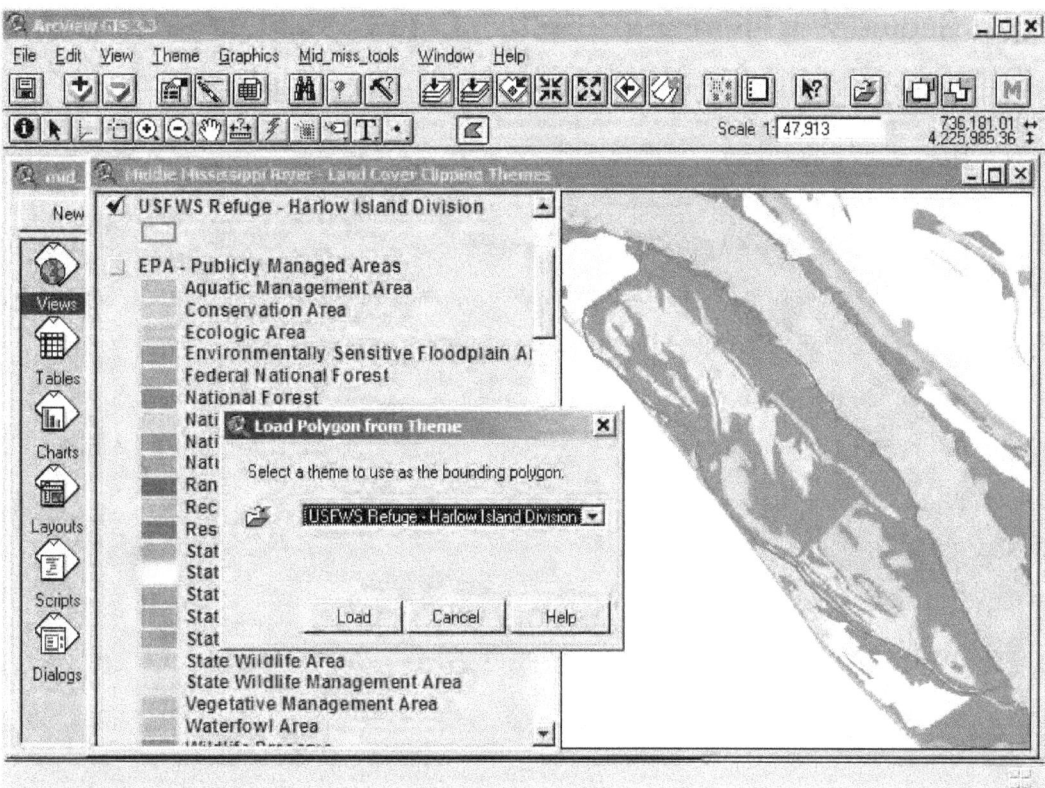

Figure 6–9.   Using a preexisting data theme to clip other data.

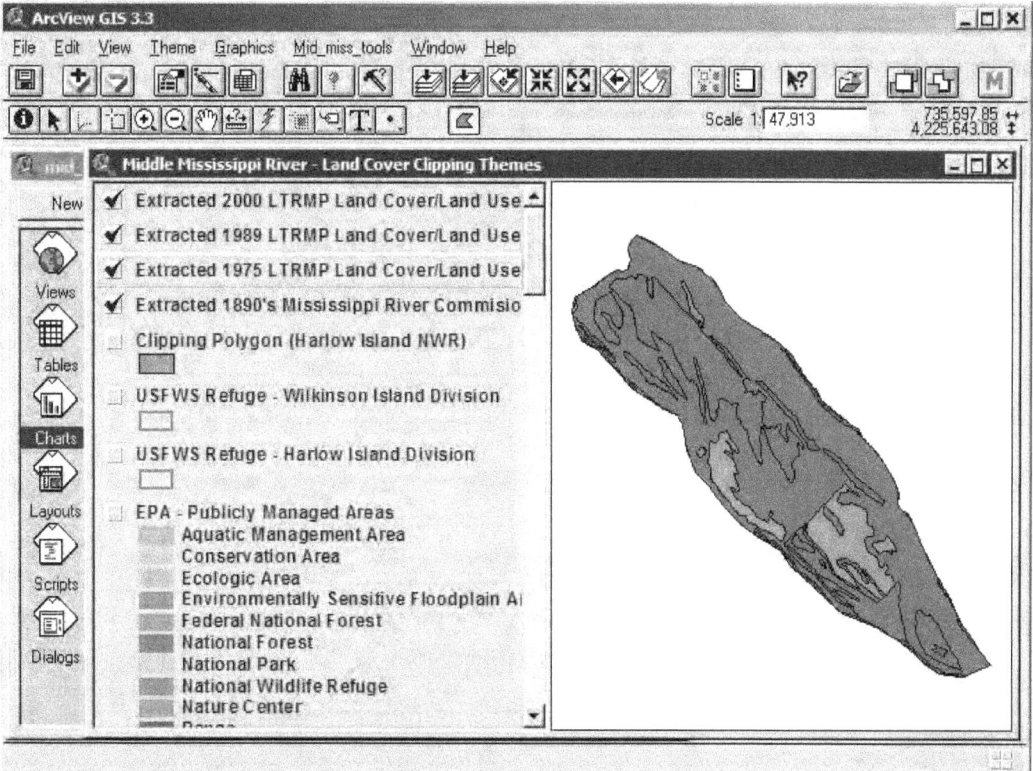

Figure 6–10.   Data extracted by using a preexisting data theme.

Appendix 2: Geographic Information System Data Theme List

The following is a list of the geographic information system data themes included with the Middle Mississippi River Decision Support System.

- Upper Midwest Environmental Sciences Center/Long Term Resources Monitoring Program (UMESC-LTRMP) Water Quality Component Data (1993 – 1999, Spring – Summer – Fall – Winter)

- UMESC-LTRMP Invertebrate Component Data (1994 – 1999)

- UMESC-LTRMP Fisheries Component Data (1993 – 1999, Spring – Summer – Fall)

- Southern Illinois University (SIU) Sturgeon Telemetry Points 1995 – 2000

- SIU Sturgeon Telemetry Points 2002 –- 2003

- Environmental Protection Agency/Upper Mississippi River Basin Association (EPA-UMRBA) Marinas

- EPA-UMRBA Dams

- EPA-UMRBA Lock and Dams

- EPA-UMRBA Boat Access Points

- Scientific Assessment and Strategy Team (SAST) River Miles

- LTRMP Feature Labels

- U.S. Geological Survey (USGS) Geographic Names Information System (GNIS) City Names

- USGS-GNIS Points

- Scientific Assessment and Strategy Team (SAST) Eagle Nest Data

- SAST Heron Rookeries

- Wing Dam Bathymetry Points

- Corps of Engineers (COE) Barge Sailing Line

- COE Barge Fleeting Areas

- EPA-UMRBA Roads

- EPA-UMRBA Railroads

- COE-Saint Louis District (STL) Weirs

- COE-STL Dikes

- COE-STL Revetments

- SAST 1993 Levee Breaks

- SAST Levee Linears

- SAST Leveed Areas

- USGS Digital Line Graph Streams

- EPA-UMRBA Streams

- USGS-LTRMP River Mile Boundaries

- LTRMP - COE 2000 Gravel Bars

- USGS-EPA - National Hydrography Dataset (NHD) Lakes

- SAST Sport Fisheries Data

- EPA-UMRBA Major Waterways

- SAST Fish Spawning Data

- SAST Commercial Fisheries Data

- SAST Mussel Beds

- United States Fish and Wildlife Service (USFWS) Refuge - Wilkinson Island Division

- USFWS Refuge - Harlow Island Division

- EPA-UMRBA Other Environmentally Sensitive Areas

- EPA-UMRBA Special Designated Areas

- EPA - Publicly Managed Areas

- LTRMP-COE Bathymetry

- COE - 2001 Aquatic Areas

- 1989 LTRMP Aquatic Area Classifications

- 2000 LTRMP Land Cover/Land Use

- 1994 COE REEGIS Land Cover/Land Use

- 1989 LTRMP Land Cover/Land Use

- 1989 LTRMP Satellite-Derived Land Cover/Land Use

- 1975 LTRMP Land Cover/Land Use

- 1890's Mississippi River Commission LCLU

- Early 1800's Government Land Office Land Cover

- 2000 LTRMP Land/Water

- UMRS – The Nature Conservancy (TNC) Priority Sites

- UMRS - TNC Ecological Drainage Units

- UMRS - TNC Catchment Areas for Headwater and Creek Systems

- UMRS - TNC Catchment Areas For Small River Systems

- UMRS - TNC Catchment Areas For Medium River Systems

- UMRS - TNC Catchment Areas For Large River Systems

- UMRS - TNC Medium River Reaches Buffered by 2 km

- UMRS - TNC Large River Reaches Buffered by 3 km

- UMRS - TNC Great River Reaches Buffered by 5 km

- UMRS - UMESC Subbasins

- UMRS - TNC Surficial Geology

- EPA Counties

- UMRS - STATSGO Soils

- COE 1998 Aerial Photos

- LTRMP 2000 True-Color Aerial Photos

- USGS Digital Orthophoto Quarter Quads

- USGS 1:24,000 Digital Raster Graphic

- USGS 1:100,000 Digital Raster Graphic

- SAST Elevation/COE Bed Elevation Mosaic

- SAST Digital Elevation Model

- USGS Digital Elevation Model

Appendix 3: FRAGSTATS Attribute List

The following is a list of landscape metric fields included within the FRAGSTATS output themes for bathymetry and aquatic habitats and also a short description of each field. Visit the Web site at http://www.umass.edu/landeco/research/fragstats/ fragstats.html for more detailed metric descriptions.

Field name	Field description
TA	Total landscape area (ha)
NP	Number of patches (#)
PD	Patch density (#/100 ha)
LPI	Largest patch index (%)
TE	Total edge length (in meters) of all selected edges (background features excluded)
ED	Edge density (m/ha) of selected features (from TE)
LSI	Landscape shape index calculated using TE (non-weighted edge length with background features excluded)
AREA_MN	Mean patch area (ha)
AREA_AM	Area-weighted mean patch area (ha)
AREA_MD	Median patch area (ha)
AREA_RA	Range in patch area (ha)
AREA_SD	Standard deviation in patch area
AREA_CV	Coefficient of variation in patch area
GYRATE_MN	Mean radius of gyration distribution (mean distance (m) between each cell in the patch and the patch centroid)
GYRATE_AM	Area-weighted mean radius of gyration distribution
GYRATE_MD	Median radius of gyration distribution
GYRATE_RA	Range in radius of gyration distribution
GYRATE_SD	Standard deviation in radius of gyration distribution
GYRATE_CV	Coefficient of variation in radius of gyration distribution
SHAPE_MN	Mean shape index distribution
SHAPE_AM	Area-weighted mean shape index distribution
SHAPE_MD	Median shape index distribution
SHAPE_RA	Range in shape index distribution
SHAPE_SD	Standard deviation in shape index distribution
SHAPE_CV	Coefficient of variation in shape index distribution
FRAC_MN	Mean fractal index distribution
FRAC_AM	Area-weighted mean fractal index distribution
FRAC_MD	Median fractal index distribution
FRAC_RA	Range in fractal index distribution
FRAC_SD	Standard deviation in fractal index distribution
FRAC_CV	Coefficient of variation in fractal index distribution
PARA_MN	Mean perimeter-area ratio distribution
PARA_AM	Area-weighted mean perimeter-area ratio distribution
PARA_MD	Median perimeter-area ratio distribution
PARA_RA	Range in perimeter-area ratio distribution
PARA_SD	Standard deviation in perimeter-area ratio distribution
PARA_CV	Coefficient of variation in perimeter-area ratio distribution
CIRCLE_MN	Mean related circumscribing circle distribution
CIRCLE_AM	Area-weighted mean related circumscribing circle distribution
CIRCLE_MD	Median related circumscribing circle distribution
CIRCLE_RA	Range in related circumscribing circle distribution
CIRCLE_SD	Standard deviation in related circumscribing circle distribution
CIRCLE_CV	Coefficient of variation in related circumscribing circle distribution

Field name	Field description
CONTIG_MN	Mean contiguity index distribution
CONTIG_AM	Area-weighted mean contiguity index distribution
CONTIG_MD	Median contiguity index distribution
CONTIG_RA	Range in contiguity index distribution
CONTIG_SD	Standard deviation in contiguity index distribution
CONTIG_CV	Coefficient of variation in contiguity index distribution
PAFRAC	Perimeter-area fractal dimension (reflects shape complexity across a range of spatial scales [patch sizes])
TCA	Total core area (ha)
NDCA	Number of disjunct core areas contained within each patch of the corresponding patch type
DCAD	Disjunct core area density
CORE_MN	Mean core area distribution
CORE_AM	Area-weighted mean core area distribution
CORE_MD	Median core area distribution
CORE_RA	Range in core area distribution
CORE_SD	Standard deviation in core area distribution
CORE_CV	Coefficient of variation in core area distribution
DCORE_MN	Mean disjunct core area distribution
DCORE_AM	Area-weighted mean disjunct core area distribution
DCORE_MD	Median disjunct core area distribution
DCORE_RA	Range in disjunct core area distribution
DCORE_SD	Standard deviation in disjunct core area distribution
DCORE_CV	Coefficient of variation in disjunct core area distribution
CAI_MN	Mean core area index distribution
CAI_AM	Area-weighted mean core area index distribution
CAI_MD	Median core area index distribution
CAI_RA	Range in core area index distribution
CAI_SD	Standard deviation in core area index distribution
CAI_CV	Coefficient of variation in core area index distribution
PROX_MN	Mean proximity index distribution
PROX_AM	Area-weighted mean proximity index distribution
PROX_MD	Median proximity index distribution
PROX_RA	Range in proximity index distribution
PROX_SD	Standard deviation in proximity index distribution
PROX_CV	Coefficient of variation in proximity index distribution
ENN_MN	Mean euclidean nearest neighbor distance distribution
ENN_AM	Area-weighted mean euclidean nearest neighbor distance distribution
ENN_MD	Median euclidean nearest neighbor distance distribution
ENN_RA	Range in euclidean nearest neighbor distance distribution
ENN_SD	Standard deviation in euclidean nearest neighbor distance distribution
ENN_CV	Coefficient of variation in euclidean nearest neighbor distance distribution
CONTAG	Contagion index (%)
PLADJ	Percentage of like adjacencies
IJI	Interspersion and juxtaposition index (%)
CONNECT	Connectance index
COHESION	Patch cohesion index
DIVISION	Landscape division index
MESH	Effective mesh size
SPLIT	Splitting Index
PR	Patch richness (#)
PRD	Patch richness density (#/100 ha)

Field name	Field description
RPR	Relative patch richness (%)
SHDI	Shannon's diversity index
SIDI	Simpson's diversity index
MSIDI	Modified Simpson's diversity index
SHEI	Shannon's evenness index
SIEI	Simpson's evenness index
MSIEI	Modified Simpson's evenness index
AI	Aggregation index

www.ingramcontent.com/pod-product-compliance
Lightning Source LLC
Chambersburg PA
CBHW081616170526
45166CB00009B/2987